The Scholarship Series in Biology

General Editor: W. H. Dowdeswell

Variation and Adaptation in Plant Species

THE SCHOLARSHIP SERIES IN BIOLOGY

Human Blood Groups and Inheritance
S. Lawler and L. J. Lawler

Woodland Ecology
E. G. Neal

New Concepts in Flowering Plant Taxonomy
J. Heslop Harrison

The Mechanism of Evolution
W. H. Dowdeswell

Heathland Ecology
C. P. Friedlander

The Organization of the Central Nervous System
C. V. Brewer

Plant Metabolism
G. A. Strafford

Comparative Physiology of Vertebrate Respiration
G. M. Hughes

Animal Body Fluids and their Regulation
A. P. M. Lockwood

Life on the Sea-shore
A. J. Southward

Chalkland Ecology
John Sankey

Ecology of Parasites
N. A. Croll

Physiological Aspects of Water and Plant Life
W. M. M. Baron

Biological Principles in Fermentation
J. G. Carr

Studies in Invertebrate Behaviour
S. M. Evans

Ecology of Refuse Tips
Arnold Darlington

Plant Growth
Michael Black and Jack Edelman

Animal Taxonomy
T. H. Savory

Ecology of Fresh Water
Alison Leadley Brown

Ecology of Estuaries
Donald S. McLusky

Variation and Adaptation in Plant Species
D. A. Jones and D. A. Wilkins

Variation and Adaptation in Plant Species

David A. Jones, M.A., D.Phil.
Lecturer in Genetics, University of Birmingham

and

Dennis A. Wilkins, B.Sc., Ph.D.
Senior Lecturer in Botany, University of Birmingham

Heinemann Educational Books Ltd
London

Heinemann Educational Books Ltd

LONDON EDINBURGH MELBOURNE TORONTO
AUCKLAND JOHANNESBURG SINGAPORE
HONG KONG IBADAN NAIROBI NEW DELHI

ISBN 0 435 61480 0

First published 1971

Frontispiece

Arum maculatum—an example of genetic polymorphism. The spotted and unspotted leaves are borne on different plants.

(Photograph by courtesy of Roy Hewson and H. A. Roberts, National Vegetable Research Station, Wellesbourne, Warwickshire)

Published by Heinemann Educational Books Ltd
48 Charles Street, London W1X 8AH

Printed in Great Britain by
Richard Clay (The Chaucer Press) Ltd, Bungay, Suffolk

Preface

The study of evolution can be approached from many different viewpoints. During the hundred years or so of its development the Theory of Natural Selection has undergone important changes of emphasis, until today it is a complex synthesis of genetical and ecological ideas. This book is not concerned with the broad sweep of evolutionary history, but with that interaction between genetics and ecology which is implicit in the concept of adaptation.

We have chosen to restrict the field in two ways. Because of our personal interests and experience we have confined our detailed comments to plants, but in spite of this we believe that most of the principles discussed are of more general application. Secondly we have concentrated on variation within species, because this is the area in which proper genetic analysis can be carried out, and because we maintain that this is where evolutionary change begins.

In the course of the book we have digressed in various ways from the narrow theme of adaptation. In particular we have included a certain amount of formal population genetics and statistics, in the belief that those readers who may be planning their own investigations will find the discussion of these matters useful. We would stress, however, that these sections are in no way essential for an understanding of the rest of the book, most of which only assumes a background of sixth-form biology.

We acknowledge, gratefully, the constructive criticism of our colleague Dr J. S. Gale and of Mr W. H. Dowdeswell, the editor of this series. We do, however, take full responsibility for the errors and omissions.

1971

D.A.J.
D.A.W.

Acknowledgements

We are indebted to the following publishers for permission to reproduce items from their publications. (1) The literary executor of the late Sir Ronald Fisher, F.R.S., to Dr Frank Yates, F.R.S. and to Oliver & Boyd Ltd., Edinburgh, for parts of Tables III, IV, V, VII, and X from their book *Statistical Tables for Biological, Agricultural, and Medical Research*. These are included here as Tables 1, 2, 3, 4, and 5. (2) Blackwell Scientific Publications, Oxford, for Figures 4.13, 4.14, and 2.2 taken from their journals *New Phytologist* and *Science Progress*. (3) Academic Press Inc. Ltd., London, for Tables 4.10 and 4.11 taken from their publication *Advances in Genetics*. The precise references are included in the legend to the tables and figures.

Contents

1

Variation

Introduction

We are all familiar with the different varieties of dogs, cattle, potatoes, dahlias, and other domesticated animals and cultivated plants. The differences between these varieties are largely the result of deliberate breeding and selection by man. The frequent differences found between individual members of plant and animal species in wild populations are less well known, yet it is upon this variation that natural selection acts. These differences are of many kinds: they can be clear-cut or continuously graded, and they may involve morphological or physiological characters.

Although it is easier to study morphological characters, and to envisage the probable effects of selection upon them, physiological characters play a highly important part in determining which plants are the more successful in a given habitat. Much of the work prior to about 1950 was concerned with showing that morphological varieties were often characteristic of different types of habitat, but more recently a great deal of physiological variation within species has been discovered and this is not always related to the more conspicuous differences of morphology.

It is likely that plants from different habitats differ in their characteristics as a result of selection, but this is not always easy to prove in an individual case. There have been several attempts to show that a single character difference reflects adaptation to a single habitat difference. Such investigations usually reveal unexpected complexities in both plant and habitat.

Variation and the species

The Cox and the Bramley both belong to the same species, the apple, and in spite of their different culinary properties they are both known in botanical circles as *Malus sylvestris*. Equally the Aberdeen Angus and the Hereford are both varieties of one species of cattle, and the Majestic and the King Edward are varieties of *Solanum tuberosum*, the cultivated potato.

Even these examples are not quite as clear-cut as they appear, however, and to avoid confusion it is important to distinguish a number of rather different situations in which the single term 'species' is used. The first is the ordinary wild species of the taxonomist. This corresponds closely with the popular idea of a 'kind' of plant or animal. The lion is one species and the tiger another. The beech and the ash are two quite distinct species of British tree. We have, on the other hand, two native oaks which are rather difficult to distinguish. It is to such wild species that the familiar Latin binomial is most commonly applied: *Quercus robur* and *Q.petraea* being the names of the two oaks.

When cultivated varieties are bred from a single identifiable wild species they retain the parental name. In the case of old established crop plants the wild parent is not always known with certainty; the example of *Malus sylvestris* mentioned above is probably of this type, and so is *Brassica oleracea*, where forms as different as the cabbage, the cauliflower, and the Brussels sprout all retain the binomial of their parent, the wild cabbage.

A large number of familiar crops on the other hand, either have no generally recognized wild relatives because of their antiquity, as in the case of the major cereals, or they have complex origins from the hybridization of a number of wild species. In these cases a new name has to be chosen which can be applied to all the varieties but which does not include any wild individuals at all. In the purely nomenclatural sense a new kind of plant has been produced.

Some of the cereals are so different from all known wild plants that new genera have been invented for them. There are no wild members of either *Zea* (maize) or *Triticum* (wheat). More often it is only a new specific epithet which is required. The great range of *Dahlia* varieties now available has been produced from the hybridization of about six wild species, some with purple and some with orange and red flowers. The varieties are all grouped together into the new species *Dahlia variabilis* which has no wild representatives.

It is clear from this discussion that species vary greatly in their homogeneity. Many wild species appear on first inspection to be completely uniform—an impression, incidentally, which this book is largely designed to dispel—while any species which includes cultivated varieties will inevitably appear much more variable. Is this an unsatisfactory state of affairs? Ought not a species to be at least reasonably uniform in its characters? This raises the whole debatable question of species criteria. The general problem will not be discussed in detail here but one or two comments are necessary because of the recurrent need to use the word 'species' in any discussion of biological variation.

The taxonomist does not pay equal attention to all characters. In any particular group of organisms he uses certain characters rather than others, basing his choice on a wide knowledge of their behaviour throughout the group and in particular on their correlation with one another. He knows for instance that flower and fruit structure allow him to make confident predictions about many other characters in most families of flowering plants. They are more informative and hence of more value in classification than leaf or stem structure. This point of view gives him a rather different picture of the variability of cultivated plants. The cabbage and Brussels sprout differ strikingly in their vegetative organization, but when they are allowed to flower we see immediately why the taxonomist says they are closely related.

Many other examples could be quoted of this general principle that domesticated varieties differ in characters which are conspicuous to the naked eye, while still preserving intact those features which identify them as members of a single species. The skeletons of a carthorse and a racehorse are remarkably similar. The obvious differences in external appearance are superficial and of no taxonomic importance.

So far we have given little indication of how the taxonomist comes to his decisions. It is also still worth considering when we ought to regard the products of the breeder as entirely new species. We have shown that *Zea* is a new genus, and *Solanum tuberosum* a new species, but that *Malus sylvestris* has both wild and cultivated members. Are there any rules by which a species can be distinguished from a variety, without having to fall back every time on the opinion of an individual taxonomist? This is equivalent to asking whether there are any characters which can always be relied upon to be of taxonomic importance.

The general answer to this is that there are no absolute rules but quite a lot of useful conventions. None of these apply equally well to mammals, grasses, and bacteria. Within the grasses any change in the structure of the flower or of the inflorescence would be treated as a change needing taxonomic recognition. *Zea mays* has a genus to itself precisely because its flower arrangement is unique.

There is one important principle of wider application than most, although it must be stressed that it is far from being absolute and may even in some situations be quite meaningless. That is the test of interfertility. There is considerable support, especially among people working with higher animals and some groups of higher plants, for treating all individuals which can be readily crossed as members of the same species, and expecting members of other species to be to some extent intersterile with them. The operation of this principle is seen in the abundance of mongrel dogs and the ready production of cross-bred sheep. Within each of these groups there is little evidence of sterility, so on this criterion we would adopt a single species name to cover each assemblage of domestic varieties. Most species of mammal, on the other hand, either cannot be crossed at all or give sterile progeny.

Unfortunately for those in search of simplicity this test is much less satisfactory when applied to plants. The hybrid origin of some varieties has already been mentioned, implying that their parental species were interfertile. On the other hand there are many examples of chromosome races within a species which cannot be crossed, and when these are morphologically indistinguishable there are objections to raising them to species level. We also find the definition troublesome when we ask how the degree of interfertility is to be determined. Two species may never hybridize in the wild and yet be freely able to do so when artificially cross-pollinated. Some of the most dramatic examples of this are found in the Orchidaceae. In this family not only species but many whole genera can be freely crossed in the glasshouse, showing that the familiar type of sterility barrier found in most other groups is largely absent. The rarity of hybrids in the wild is due to a very complex pollination mechanism, which ensures that pollen from one species is distributed by insects to other members of the same species only. The breeder, however, is free to transfer pollen where he wishes, by using a needle or a brush instead of an insect, and so can obtain a host of hybrids which could not arise in the wild.

Origin and types of variation

The basic similarity between parent and offspring is due to heredity, but there are also differences between them and these may or may not be inherited. These differences constitute variation. There is a relation between the genotype and the phenotype which is dependent upon both the nature of the genotype and the environment in which development takes place. Thus two individuals with identical genotype will show differences when they are allowed to mature in different environments.

Non-inherited variation—Phenotypic plasticity

Thus it is not correct to assume that the effect of the environment is merely to determine success or failure. An additional effect, common in plants, is the ability of the individual to change the developmental course of some characters in response to environmental pressure. This can be seen in several different ways.

If a single plant is grown for a period in a warm glasshouse and then moved to the cooler environment outside it will, if it survives the move, show some morphological changes. Most probably the leaves it produces out of doors will be smaller than those it produces in the warmer and darker conditions of the glasshouse. Its new leaves are likely to be paler green, and the internodes of the new stems shorter. The genotype of the plant has not changed. The modifications to the phenotype are purely a result of the individuals plastic response to a change in the environment, an ability which is itself genetically determined.

This implies that when we compare two plants we cannot tell at sight whether the differences between them are due to genetical differences or to their having been grown under different conditions. Two tomato plants of different heights, for instance, may either belong to a tall and a dwarf strain, or they may have been grown in soils of markedly differing fertility. In a great many practical cases we have to deal with combinations of these two causes of phenotypic differences. The importance of phenotypic plasticity becomes obvious when we wish to deduce something about a plant's genotype from its external appearance.

Inherited variation

Although there is remarkable stability in the chemical structure of the nucleic acids which make up the genotype, recurrent changes can and do occur affecting individual genes, whole chromosomes or parts of chromosomes. These changes of genotype are mutations but they are not the only raw material for evolutionary processes nor the only way by which variation arises.

The other method involves the reassortment of existing mutations into new combinations. This comes about during meiosis. Firstly there is the segregation of whole chromosomes and the genes which they carry at Anaphase I. In addition, there is the major consequence of chiasma formation during pachytene whereby segments of chromosome may be exchanged between homologous chromosomes. Furthermore, there is the reduction in the chromosome number from the diploid to the haploid state. By these means new combinations of existing genes can be formed so that, at a subsequent fertilization, a zygote with an unique genotype will result. Segregation is only effective with heterozygous loci, but except in a few habitually inbreeding plant species it is usual for sexually reproducing organisms to be heterozygous at very many loci.

Unfortunately, mutants are being lost by chance as a result of meiosis and fertilization merely because they do not happen to enter a zygote. Mutation, however, can occur at any time and in any cell and so the pool of mutants is being constantly topped up with new ones.

Table 1.1 shows the four types of mutation which can be recognized. Inversion involves a change in sequence of the genes on a chromosome as a result of a segment of the chromosomes being inserted the 'wrong' way round. Translocation is the attachment of a chromosome segment to a non-homologous chromosome. A special case is the interchange when non-homologous chromosomes have exchanged segments. These aberrations may come about as errors of the meiotic process but we still need to know more about their origins.

Table 1.1

The four classes of mutation which can be recognized.

Mutation of	Type of mutation	Examples
Genes	Qualitative	Change in chemical structure of the gene
	Quantitative	Duplication or deletion of genes from the chromosomes
Chromosomes	Qualitative	Inversion, translocation
	Quantitative	Polyploidy (addition of whole chromosome sets) Aneuploidy (addition or deletion of whole individual chromosomes)

Various examples of these types of mutation will be discussed elsewhere but for the moment we will look at the effect of qualitative mutation in genes of diploid organisms. If a new mutant gene is dominant its effect will be expressed immediately in the phenotype of the individual possessing it, provided of course that the new mutation occurs before the gene is due to act and is in the appropriate tissue. A gene determining flower colour which mutates in the root can have no effect on the colour of the flowers borne by the stems. In addition if the gene mutates during the formation of a stem then only the cell descendants of the mutant cell will contain the mutation, and so only the flowers borne on that stem will show the mutation. A mutation is of importance in the long term only if it occurs in the cell line ancestral to the formation of embryo sacs or pollen mother cells. Study of mutation processes and their consequences has confirmed the suggestion that no gene acts by itself, but only as an integrated part of an extremely complex metabolizing system with an even more complex controlling mechanism. In this way genes can be turned on and off according to the chemical environment in which they are found. Indeed there are some genes whose sole function appears to be the control of those other genes whose job it is to produce structural or enzymatic proteins.

One of the best documented examples of recurrent mutation is that of Achrondroplasia in man, where the condition is due to a single dominant gene. Achrondroplastic dwarfs have almost normal bodies but very short arms and legs. The reproductive capacity of the dwarfs is so reduced that for every hundred children born to their normal brothers and sisters the dwarfs have only twenty. The mutant gene is, therefore, being removed from the population by selection at a rate of 80 per cent per generation. Rather surprisingly Mørch (1941) has shown that 80 per cent of the Achrondroplastic dwarfs born in Denmark possess a newly arisen mutant gene and hence the 80 per cent elimination by selection is precisely balanced by the rate at which new mutant genes are appearing in the population. The proportion of Achrondroplastic dwarfs in Denmark remains almost constant at about 33 per million.

All newly arisen mutations will appear first in the heterozygous state and therefore a new recessive in a diploid organism cannot have any effect on the phenotype. Indeed, such a mutant may not be detected at all unless the organism is self-fertile or the gene spreads to such an extent that there is the possibility of mating between heterozygotes. For this reason it is difficult to find recessive mutants in wild populations, but in cultivated plants several are known. In the tobacco plant *Nicotiana tabaccum* a mutation in a gene affecting the production of chlorophyll sometimes occurs. Heterozygotes are indistinguishable from the normal homozygotes but the progeny produced by selfing heterozygotes contain one quarter albinos. The albino plants without chlorophyll die as seedlings.

Some new mutations may have an intermediate effect on the phenotype of the heterozygote, but the consequences of this will be discussed later. Even though mutation is a recurrent phenomenon, recent work on the fine structure of genes in bacteriophage suggests that subsequent mutations in the same gene usually do not occur in the same chemical grouping. In man, mutations occur with such frequency that Stern has estimated that each human zygote contains on average four newly arisen mutant genes.

The mutation rate of any particular gene has not been adequately defined but something of the order of one mutation at a given locus per million cell generations is taken as an average rate. There are other ways of expressing the units but it remains difficult to obtain satisfactory estimates of mutation rates.

Having discussed briefly the origin of new variation we will now turn to the types of variation which can be recognized.

Continuous variation

(*a*) If we consider a sample of men with height ranging from 1.525 m (5 ft) to 1.825 m (6 ft) and measure their heights to the nearest 0.05 m, there will be six classes. On the other hand, when we measure to the nearest 0.01 m, we obtain five-times as many classes. The limit to the number of classes depends upon the accuracy with which we care to measure height. In fact the number of classes is arbitrary; it is determined by the observer and not by the character itself. The characters of height in man or yield per plant in cereal crops show *continuous* variation.

We can take this argument further. When the height of a man is measured a single metrical character is recorded yet this character is the sum of the length of the leg, the length of the body, the length of the neck, and the length of the head, each being under the control of a different group of genes but all eventually contributing to the single character of height.

Furthermore, if we now examine only those men who are 1.80 m high, some will have long bodies and short legs, others will have short bodies and long legs while others will be of intermediate structure. It is possible, therefore, for a metrical character to be determined in a number of different ways and so it seems that several different gene combinations may have the same net effect on the phenotype.

(*b*) There is one type of continuous variation where the character is scored as a whole number of indivisible units. Examples of these meristic characters are the number of petals per flower in *Ranunculus ficaria* (celandine), leaflet number per leaf in several species of *Vicia* (vetches) and the number of bristles on a particular segment of the leg of *Drosophila melanogaster* (fruit fly). Although, occasionally, a flower may have half a petal, in practice this half petal is counted as one petal for it is clearly neither a sepal nor a stamen. This type of variation may be treated as continuous because the classes are not distinct states, but merely a sequence of numbers, and it is perfectly legitimate to handle them in the usual arithmetic way to calculate, for example, mean values.

Discontinuous variation

Two distinct classes are specified when the character sex is measured in man. An individual is either male or female, hermaphrodites do occur, but they are very rare. Here the number of classes is determined by the character itself and not by the means used to measure the character. The system can be compared with a switch with two positions, one for male and the other for female. This type of variation is described as *discontinuous*. Other examples may be less clear-cut than this. The occasional albino seedling of *Nicotiana tabaccum* shows a complete lack of chlorophyll. Flower colour in *Lathyhus odoratus* (sweet peas) depends upon different proportions of pigments, and yet the overall phenotype can be treated in a discontinuous manner. An unusual example is offered by the tall and short varieties of garden pea used by Mendel in his classic experiments. Plant height normally shows continuous variation, but these two particular varieties were sharply separated and even in crossing no intermediates appeared.

Two distinct categories of discontinuous variation have been defined.

(*a*) Discontinuous variation may be maintained by a balance between selection and recurrent mutation. The example of Achrondroplasia given earlier fits here, and so do the chlorophyll mutants of *Nicotiana tabaccum* and other plants.

(*b*) Some forms of discontinuous variation cannot be maintained merely by recurrent mutation and this has been defined in a formal manner by Ford (1940). *Genetic polymorphism* is the occurrence together in the same interbreeding population of two or more forms of the same species in such proportions that the rarest of them cannot be maintained merely by recurrent mutation. This implies that selective forces are in balance and this point will be expanded later. Polymorphism is widespread in animals and has now been found to be much commoner in plants than was originally thought. Flower colour in *Iberis amara* (candytuft) where the plants have either white or mauve flowers, but not both, is a good example. Similarly in wild populations of *Lotus corniculatus* (birds-foot trefoil) some plants have yellow and others brown-tipped keel petals. Further examples are flower colour in *Polygala vulgaris* (milkwort), distyly in *Primula* spp., and incompatibility systems in *Prunus* (cherry) and *Oenothera* (evening primrose) species.

Artificial selection

The main concern of this book is with wild plants, but there are still one or two further points which can be better illustrated from cultivated material. The differences between varieties (or cultivars, as the plant breeders have agreed to call them) have arisen as a result of deliberate choice on the part of the breeder. Sometimes all that has been done is to select the most suitable individuals from the wild to act as breeding parents for the production of commercial seed. Some 'species' in the garden catalogues are of this kind: little changed from their wild condition. This simple situation does not last long. In each cultivated generation the breeder naturally selects the best from which to propagate, and all the older cultivars and animal breeds were produced by the prolonged application of this process for generation after generation. A surprising amount can obviously be done by simple but progressive selection, without any conscious application of genetical principles at all.

It is true that more rapid progress has been possible since the development of genetical theory allowed crossing and selection to be put on a firmer basis, but the important point here is to stress the similarity between the artificial selection exerted by the breeder and the natural selection which leads to diversity in the wild. We must not, of course, ignore the conspicuous differences between the two processes, especially in the type of character selected. The breeder selects for improvement in a small number of chosen characters which are of direct benefit to him, whether this involves milk yield in cows, protein yield in fodder grasses, disease resistance in potatoes, or colour and size in roses. Sometimes his aims seem irrational or even frivolous, as in the production of ever more bizarre types of dog. Sometimes unexpected irrationality creeps into the seemingly solemn matter of food production, as in the case of the potato breeder who reputedly has to ensure that his new variety has pink eyes because the housewife regards this as the hallmark of quality in potatoes. Commercial varieties have, after all, to be sold.

There are many other interesting problems facing the breeder. He may have to reconcile contradictory requirements. In many vegetables, for instance, the consumer requires a leafy plant which does not waste its substance by producing flowers. On the other hand, because these plants must be grown from seed, the merchant has to be able to produce seed cheaply and in large quantities. This is achieved in the case of the lettuce by selecting for very late flowering. There is then a long period of vegetative growth during which a large leafy plant can be harvested if required, but this is followed late in the season by free flowering and the production of ample seed. If selection was for vegetative characters alone there is no doubt that flowering capacity would be reduced to uneconomic levels.

Modern agricultural methods also impose unexpected requirements. Sugar beet is sown as a 'seed' which is really a cluster of four fruits fused into an irregular ball. This is too uneven in size and shape for efficient mechanical sowing, and gives rise to uneven spacing between plants. A big effort is being made to perfect a 'single-seeded' beet which should be more satisfactory in both respects.

In addition to these specific aims the breeder must also ensure the general health of his products. They must be vigorous and able to thrive under the type of management for which they are intended. This is a rather more subtle matter than it appears. Modern varieties only thrive under modern management. They are produced for the specific purpose of giving maximum performance under intensive care, such as high levels of fertilizing or feeding, and they may be inferior to older varieties under more primitive conditions. This highlights an important fact about adaptation which we shall emphasize repeatedly. There is no such thing as generalized fitness for all environments. A variety which does well under one set of conditions may do very badly under another. This is the principle underlying the current development of special-purpose strains which will do extremely well under closely specified conditions of management, and it applies to a whole range of examples from malting barley suited to the Scottish climate to broiler chickens for battery rearing.

Competition and survival

Cultivated varieties lead a sheltered life by comparison with their wild relatives. The wild plant has to establish itself unaided, and then survive to maturity and reproduce effectively, under conditions of severe competition from its neighbours. The domesticated plant, on the other hand, is normally protected from competition, especially at the seedling stage, and is very often not required to reproduce at all. Many crop plants, such as strawberries and potatoes, are propagated vegetatively in order to preserve them as pure strains. They are genetically heterogeneous and sexual reproduction would lead to segregation and the complete loss of the very complex of characters which makes the variety what it is. Once sexual reproduction is no longer demanded as a condition of success there is no reason why completely sterile forms should not be as good as any others, and in fact many familiar varieties never set seed at all. The Bramley Seedling cooking apple, for instance, has three sets of chromosomes per cell, in other words it is a triploid, and the resulting breakdown in meiosis makes the plant virtually sterile. It is none the less a highly successful crop plant, and it is easily propagated by grafting. The seedlessness of the common banana is even more conspicuous.

This is in marked contrast to the situation among higher plants in the wild, where vegetative reproduction is usually no more than a supplement to sexuality, and where sterile individuals are uncommon and rarely persist.

The protection of cultivated plants from competition is familiar to everyone. Many of the traditional agronomic practices found in any farming community are directed to the elimination of weeds or pests or to the regulation of the density of the crop itself. The whole subject of weed control has recently been drastically altered by the success of hormone weedkillers, which allow various groups of unwanted species to be poisoned selectively in a highly specific way. This has resulted in the virtual elimination of many plants which were once serious pests of arable crops. The competitive effects of neighbouring plants of the crop itself are perhaps less obvious. The farmer recognizes that there is an optimum seeding rate for any crop. Too small an amount of seed clearly gives a low yield. As the amount of seed is increased so the yield per hectare increases, but only up to a certain point. The effect of competition is to make the individual plants smaller and at some stage this is severe enough to balance the increasing number of plants resulting from high seeding rates. From this point onward there is no gain of yield in return for the increasing cost of more seed. It is even possible to get a reduction in yield with too high a density. This is true of any field crop, including forest trees, which have to be rigorously thinned for success, and even of grazing animals, where too high a density can again lead to a reduced total yield of meat or milk.

Wild plants have none of this protection from the competitive effects of their neighbours. Very few of them grow in open communities where the plants are too far apart to influence each other. They are exposed to intense competition from the moment of germination of the seed, and we can be sure that those which survive to reproductive maturity are the ones most capable of meeting these pressures. We shall return later to discuss the mechanisms of competition among wild plants and its bearing on natural selection. The point to emphasize here is that cultivated plants have lost some of the competitive vigour possessed by their wild ancestors, simply because they have been selected for success under less demanding conditions. The truth of this statement is demonstrated by the widespread failure of cultivated plants to establish themselves in wild communities. There are abundant opportunities for escape, but few can survive the rigours of life in the wild.

Adaptation

This term has already been used several times in what were intended to be non-controversial ways. It is nevertheless an ambiguous word in biology and has been the centre of a good deal of argument. Much of the trouble arises from the fact that a word in common use has been taken over by science and given a technical meaning. As always when this is done the familiar overtones of the word persist and tend to confuse any precise technical definition.

The everyday sense of this particular word carries a suggestion of deliberate planning. An old tool is adapted to a new use by altering it slightly so that it can be used in a different way. The change is made by a human craftsman after careful thought, involving the comparison of its old and new uses and the planning of the most appropriate alterations. It is extremely difficult to import such a word into biology without at the same time preserving in many people's minds the idea of deliberately making a change for a purpose. Few biologists would in fact wish to imply that adaptation was the result of a design imposed from outside, but unfortunately the very use of the word 'adaptation' tends to do just that. It may be argued that the wrong word has been chosen altogether, but it is too late to change it completely as its use is too widespread. The best we can do is to try and define the technical meaning we intend and then stick to it. This is a widespread difficulty in scientific writing where it is not always possible to invent completely new and neutral terms for ideas which are themselves either new or newly defined in a more rigorous way. It has been argued that Natural Selection is neither natural nor selection, but it remains an indispensable technical term whose meaning is reasonably widely understood.

The external difference between the two situations, to revert to our misleading analogy of the tool for a moment, is that the craftsman can tell us why he adapted his tool in the way he did and it makes sense to ask him why. In biology we merely have two plants with different characteristics and which are found in different environments, and there is no one to ask.

We carry out experiments on our two plants which are designed to test their fitness in various environments, and then we make deductions about the differences between them and the way in which these differences allow them to grow and reproduce in a range of natural environments. We can investigate the physiological mechanisms which enable one plant to grow better than another, and we can study the genetic basis of the difference and the way it is inherited. None of these activities involves us in any assumptions about purpose or design, and yet we feel fully justified in saying we are studying the adaptation of the plant to life under different conditions.

It may help to distinguish two possible usages. We can talk about the process of adaptation, by which a plant changes in the course of time so that it becomes better fitted to a particular environment. This is close to the vernacular sense of the word, but is rarely met with in botanical writing. It is much more usual to discuss the degree to which a plant is adapted to various habitats at a single point in time, or to compare the adaptive properties of a number of plants. Adaptation is then not a process but a state, and it is in that sense that we shall use the word in this book.

There is one type of biological adaptation, all the same, which does in some ways approach the first meaning given above. If an animal experiences a change in its environment it can often modify its behaviour to take account of the change—for instance a change in available food may lead to a change in feeding habits. This may seem so trivial that it hardly merits a technical term at all. The parallel situation in plants, however, is instructive. There is little scope for an alteration of behaviour in response to a change of conditions, and a plant normally responds by changing its morphology, as in the examples of plasticity already touched on. This requires growth and is consequently slow, but it clearly involves the modification of an individual during its lifetime in such a way as to change some of its properties, and so may be thought of as an example of adaptation as a process. The danger lies in assuming that any plastic change is automatically to be regarded as an adaptation. As will be discussed elsewhere many such changes are adaptive, but it is misleadingly facile to treat the two as synonymous. There is no reason in principle why non-adaptive plastic changes should not be induced in a plant by suitably manipulating its environment.

There is another term which is frequently used as an alternative to adaptation, particularly in discussions of genetics and selection, and that is 'fitness'. In many contexts the two words mean the same thing, but the emphasis is different. Fitness is sometimes discussed as if it were relatively independent of the environment, whereas when we talk about adaptation we always specify the type of environment we have in mind. In fact, of course, generalized fitness is an illusion. No organism is equally fit in all environments, although admittedly one may thrive in a greater range of environments than another and so seem on average to be more generally fit. Even in this case the conclusion depends on the actual nature of the environments in which the organisms were tested.

Most of the differences between individuals and populations we wish to describe as adaptive are of genetic origin. Plasticity within a single genetic individual will be discussed in Chapter 3, but it is not our main concern.

The general question of experimental evidence must now be considered. What can be learnt about adaptation from experiment, and are there many different kinds of experiment yielding different kinds of evidence? There is an important preliminary to clarify first. In all such experiments we are making comparisons—between individuals, populations, habitats, and so on. It follows that our conclusions are always expressed in comparative terms. If a statistical analysis is employed to evaluate the results our conclusions will usually involve the statement that two populations, for instance, are significantly different in respect of a particular character. If all our evidence is of this type we ought to be wary of statements about adaptation which do not include an explicit comparison. Such statements are freely made. Surely an orchid flower is adapted to insect pollination, and a giraffe's neck to browsing from trees? We would argue that even in such cases a comparison was in fact implied, even though it was so obvious as not to be worth stating. Long necks are only adaptations to browsing if some animals have short necks, or at least if short-necked animals can be reasonably postulated. Unless we insist on this comparative approach it is easy to make meaningless statements. All plants are adapted to their environments—of course, or they would not survive, but that is not a statement which can be tested by experiment or observation, and in this sense it has no scientific meaning. The design of experiments consists largely in the choice of meaningful comparisons.

In this context it is obvious that there are only two primary experimental designs. We can either ask which of two individuals (or populations) is better adapted to one particular environment, or we can compare the adaptation of one individual (or population) to a series of different environments. The need for a criterion of success immediately presents itself. In each of these experiments we have to decide what to measure as evidence of adaptation. There seems to be no single complete answer to this. If we are interested in long-term evolution we must adopt the difficult but compelling Thoday definition (1953)—'the probability of leaving descendants after a given long period of time,' by which he implies a large number of generations. In practice we have to make an approximation to this and merely compare reproductive ability—usually in only one generation. There are a few situations, on the other hand, where we may be interested in short-term advantage rather than in evolution, and we may then regard vegetative success as more important than reproduction.

The two types of experiment are not identical in the problems they pose. When two or more population samples are grown together in a test environment it is usually not difficult to say which is the most successful. In this case the problems only arise when we wish to extrapolate to wild conditions, and make deductions from the experiment about what is likely to happen in a particular natural environment. The difficulties with the other type of experiment are more subtle. When we grow samples from a single population in different environments we have to make measurements of their performance in each environment and then go on to make deductions about adaptation from these. This often amounts to a search for an optimum environment for our population. In the simplest examples a series of conditions differing by a single factor only is tried—a species of *Sphagnum* (bog moss) may be grown in culture at many pH levels, for instance. A suitable measure of performance then gives an indication of the range of tolerance of our sample to pH. We can pick out an optimum value to which it appears most highly adapted, and in some cases extreme maximum and minimum values for growth. The difficulty again lies in extrapolating from the experiment to the wild.

We may find an optimum pH of 4.5. If other species could be ignored it would be reasonable to expect to find our test species thriving best on peat of about that pH. The snag lies in the competitive effects of other species.

There may well be another species of *Sphagnum* which grows even more vigorously at pH 4.5, but which does very badly at pH 4.0. In that case our species may be found chiefly at the lower figure, rather than at the value indicated as the optimum in our experiment. No experiments which ignore competition can be safely used to predict what we will find in the wild, but in a case like this we should obviously have learnt a great deal more by testing all the available species of *Sphagnum* in our range of culture solutions. It is very often the case that the two simple experiments have to be combined if useful information about adaptation is to be obtained. We do not compare populations or environments as in the simple scheme suggested at first, but we try to make our experiment comprehensive enough to include both comparisons.

To summarize: a biologist when faced with this statement—'this plant is adapted to its habitat' would argue thus. Of all the plants which could grow in the habitat the ones which do are those which possess characters enabling them to make the most efficient use of the resources available. These plants grow or perhaps reproduce most efficiently. The characters concerned do not arise in direct response to the habitat conditions. They arise from the complex of mutation, random assortment of whole chromosomes and recombination within chromosomes (crossing over) during meiosis, and the fusion of gametes of diverse genotypes at fertilization. The part played by the environment is the sifting out of the variation so that the well-adapted individuals are more successful than the others and so contribute more to successive generations.

This, in brief, is the theoretical background upon which discussions of variation in natural population must be based. When we set out to study this variation directly, we must consider many possible approaches, some of which are reflected in the following questions:

1. What is the spatial distribution of the variation?
2. What is the genetic nature of the variation?
3. What are the developmental mechanisms in the plant which bring about the phenotypic differences, and how subject are they to environmental modification?
4. How do the genes act to initiate these developmental pathways?
5. What selective agents have been involved in the evolution of the present populations? How have these agents acted and how strong has been the reaction to their influence? Are different stages in the life-cycle of the plant selected in different ways?

These questions have to be tackled by a number of fundamentally different techniques, including those of genetics, biochemistry, and ecology. We will now attempt to draw some of these diverse threads together, and to outline a few of the methods and ideas involved in the study of micro-evolution in plants.

2

How is variation measured?

Methods of sampling populations

Lotus corniculatus L. shows a polymorphism for keel colour in wild populations, in that some plants have yellow and others have brown-tipped keel petals. Suppose we come across such a population: how do we set about determining the frequency of the dark-keeled form? It is obvious that the best way is to count all the plants in the population. This presupposes (1) that all the plants are in flower at the same time, and (2) that it is possible to recognize the geographical limits and hence the size of the population. The first objection may be met by labelling all the plants which have not yet flowered, but there is the obvious danger that labels may be lost before the recording is complete. The second difficulty is usually ignored; the reasons are implicit in the following discussion.

If the population is large much labour is needed to find the frequency of dark-keeled plants by complete enumeration. The way round this difficulty is to take a sample and then use the frequency of the dark form in the sample as an *estimate* of its frequency in the population as a whole. It can be shown that examining a large sample is almost as good as studying the whole population, but in the above example we have to assume that the plants which have not yet flowered contain the same proportion of dark-keeled individuals as do those already in flower.

There are various ways of collecting a suitable sample. The most satisfactory estimate of the frequency of the dark-keeled form is obtained by ensuring that any one plant has the same chance as any other of being included in the sample. That is, a random sample is taken of the population. It is, unfortunately, very much easier to say 'take a random sample' than it is to collect one. It is virtually impossible to stand in front of an array of growing plants and select a number of individuals at random in the strict sense of the word. We have to make do with other methods and the one which approaches this ideal is to set out a narrow belt transect and sample every (say) tenth plant along it, starting from one end. A series of such belts should be set out to cover as much of the areas as possible. This systematic sampling method only works with species whose individuals can be clearly identified.

A more usual method is to select random points instead of random plants in the area in which they are growing and take the nearest plant to each point. If we are to do this the points must be spaced well apart with respect to the distances between the plants, or one individual may be near a number of our samples points and be counted each time. Units of 5 cm might do for small plants, but 10 metres or more would be needed for trees.

The time-honoured method of obtaining random samples of vegetation for ecological work is to throw the quadrat into the air without looking and let it lie where it falls. The equivalent in our case would be to toss a marker of some kind and take the plant nearest to it, or for small plants to toss a quadrat and take the plant nearest to the centre. Such samples are never truly random, but may be sufficiently good for a number of purposes.

Another type of non-random sampling is often used because it is easy and quick. If we put down a grid and sample the nearest plant to each intersection, or several parallel lines and take samples at fixed distances along them, we are collecting what are from an ecological point of view systematic samples (see Greig-Smith, 1964) in that they are arranged in a regular pattern over the area. As samples of vegetation their use is limited because of this. When we use the same method for sampling plants of a species, on the other hand, this may not be important, as the genetic variation is unlikely to be distributed in a regular small-scale pattern. In very many cases we would expect systematic sampling of this kind to produce a random sample of the genetic variation within any one species.

The snag about non-random sampling is that although we can use it to estimate the mean value of a character, we do not know how accurate that estimate is. This is particularly important when we wish to compare the means from two different areas.

There is another aspect of sampling which is particularly relevant to the study of continuous variation. In the example of *Lotus corniculatus* we were concerned with the relative frequencies of two forms which could be readily distinguished in the field, and the character of keel colour was one not much influenced by ordinary changes in the environment. If, on the other hand, we want to compare the expression of a continuously varying character in two populations it is usually necessary to grow samples of each under the same environmental conditions, because of the great plasticity often shown by such characters. In another instance we may wish to investigate the mode of inheritance of a character. In either case we have to collect our sample from the wild and grow it in a glasshouse or in an experimental field.

What do we sample?

There are essentially three stages during the life-cycle of seed-producing plants which can be sampled: seeds, seedlings, and mature plants, which correspond with the potential, the viable, and the established respectively. The majority of annual plants do not transplant well so it is usual to collect seeds from them. With biennials it is again probably better to collect seed although, if flowering time is the character to be studied, young plants in their first summer could be dug up, transplated and scored for flowering the following summer. When the study involves perennials we have a choice of seeds, seedlings, or mature plants, and which stage is collected depends upon the detailed aims of the work.

The breeding system and length of life-cycle must be taken into account when choosing the species to use for a particular study. For example, it may be uneconomic to sample seeds of trees unless the project is planned to last for several years. Indeed, if the choice is open, it is sensible to select plants which are annuals or if perennial, species which flower in their first year. This is related to the reasonable desire for obtaining the maximum amount of information with the minimum amount of effort in the minimum time.

Assuming a perennial species has been chosen and that it can be obtained and cultivated at any stage from seed to maturity, what criteria are involved in the decision as to which stage to collect? If we sample a mature plant we are collecting a genotype which has proved itself in the habitat. It has successfully established and developed, possibly over a long period, and its survival shows that it is suited to the conditions. If it is an outbreeding plant and we collect seed from it, on the other hand, we are sampling the material from which the next generation will be selected. Plants produce a great surplus of seed, and the few which survive to maturity are likely to be those best suited to the particular conditions, so that our seed sample may well contain individuals which would never have survived had they been left to their fate in the wild. This is particularly serious in an outbreeding plant because of the danger that it will have been pollinated by another plant growing some distance away under perhaps quite different conditions. In such a case a seed sample may contain a high proportion of poorly adapted individuals. The popularity of perennial grasses for experimental cultivation may be due to the ease with which both seed and vegetative samples can be obtained and cultivated. Particularly informative results can be obtained with such material from a comparison of the two types of sample as will be seen in Chapter 4.

Analysis of data

For the rest of this chapter we will be concerned with some of the statistical methods used to analyse biological data. We have chosen to describe those tests which are of the most use in the study of wild plants in their natural habitats, so this chapter is not in any respect intended to be exhaustive, but it is hoped that the techniques described will encourage interest in their limitations as well as in their power and thus stimulate further reading.

If several samples have been obtained by one of the methods described above it is then possible to compare them using statistical methods. We have to obtain criteria which will show whether differences between these samples are of statistical significance, or whether they are no greater than might be expected to arise from chance fluctuations of random sampling. The statistical methods to be described below contain such criteria.

Analysis of discontinuous variation

One of the first points which usually becomes clear early on in the investigation is whether the character being studied shows continuous or discontinous variation. In *Lotus corniculatus*, keel colour shows discontinuous variation in that the keels are either brown or yellow. Other characters such as flowering time, plant height, plant weight or yield of fruit or seeds are continuously variable.

The statistical tests used to determine whether two samples differ from each other depend upon the type of character studied. The χ^2 (Chi squared) test is usually sufficient for discontinuous characters while the Analysis of Variance is necessary for continuous variation. Let us begin with discontinuous variation.

The test of significance

We have taken *Lotus corniculatus* as an example up to now so let us suppose that we obtained a sample of 118 dark- and 314 light-keeled plants from population *A* and 79 dark and 336 light from population *B*. Do the populations differ in the frequency of the dark form?

If we now construct a table we have:

	Dark	Light
Sample *A*	118	314
Sample *B*	79	336

This is a 2×2 table and the test to use is the 2×2 contingency χ^2 test. Fill in the marginal totals:

	Dark	Light	Totals
Sample *A*	118	314	432
Sample *B*	79	336	415
Totals	197	650	847

This table can be generalized algebraically as:

a	b	$a + b$
c	d	$c + d$
$a + c$	$b + d$	$a + b + c + d = N$

Which gives

$$\chi_1^2 = \frac{(ad - bc)^2 \times N}{(a + b)(c + d)(a + c)(b + d)}$$

What we are doing is testing whether the ratios $\frac{a}{b}$ and $\frac{c}{d}$ are equal.

If $\frac{a}{b}$ does equal $\frac{c}{d}$ then $ad - bc = 0$ and $\chi^2 = 0$. The greater the difference between *ad* and *bc* the larger becomes χ^2 and therefore the size of χ^2 can be used as a means of measuring the difference between the two products. There is a table in Appendix III showing the probability that the value of χ^2 will be as large or larger than the value we obtain in the test. The tables allow for the circumstance when although *ad* should be the same as *bc*, we actually obtain values of *ad* and *bc* which differ by chance, i.e. because the samples are not exactly representative of their respective population.

Putting this another way; we suggest the hypothesis that populations *A* and *B* do not differ from each other in the frequency of the dark form. What is the probability that we would get two samples which differ to the observed extent merely by errors of sampling?

In the present example $\chi_1^2 = 8.13$. The subscript to the χ^2 is the number of degrees of freedom (see Appendix I for an explanation of this term and for levels of significance). Therefore by reference to the tables of χ^2 it will be seen that the corresponding probability is less than 1 per cent (i.e. $P < 0.01$) and thus we would be justified in rejecting the hypothesis that the two populations have the same frequency of the dark-keeled plants.

This is all that the test can tell us. It does not tell us *why* the populations differ, only that they *do* differ. Further work is required to explain the difference. Not only can you use the test to distinguish between two populations, but as a check on sampling techniques you can take two independent samples from the same population and test these against each other.

The form of the χ^2 test given above is not applicable when we are interested in comparing observed results with those expected on some prior hypothesis. Let us now consider a situation of common occurrence when studying populations of animals and plants even though its relevance will not be explained until the next chapter. Suppose we find a population of 450 plants in which 307 have spotted leaves and the rest plain leaves. Formal genetic analysis shows that spotted leaves are dominant and that 254 of the plants are heterozygous. We can therefore write the numbers of each genotype as:

SS	*Ss*	*ss*	Total
53	254	143	450

On counting we find that there are 360 *S* alleles and 540 *s* alleles, so the sample frequency of *S* is 0.4 and of *s* is 0.6, there being 900 alleles altogether.

It will be shown in the next chapter that the expected frequencies of the three genotypes, assuming no selection, no migration, and no mutation are $p^2:2pq:q^2$, where p and q are the population frequencies of the dominant and the recessive alleles respectively. In our case we have the estimates $p = 0.4$ and $q = 0.6$. Therefore the expected numbers are $450 \times p^2$ of *SS*, $450 \times 2pq$ of *Ss*, and $450 \times q^2$ of *ss*, i.e. 72, 216, and 162 respectively.

The formula of χ^2 we now need is

$$\chi^2_n = \sum \frac{(\text{Observed} - \text{Expected})^2}{\text{Expected}}$$

i.e. the sum of (the observed minus the expected numbers squared divided by the expected number).

Rearranging the table above to aid computation.

Genotype	Observed	Expected	O − E	$(O - E)^2/E$
SS	53	72	−19	5.01
Ss	254	216	+38	6.69
ss	143	162	−19	2.23
Totals	450	450	0	13.93

There are three classes here and so we would expect the chi-square to have two degrees of freedom. We have, however, calculated the numbers expected from the allele frequencies *estimated* from the data. As we have used one degree of freedom to do this, the chi-square we compute has only one degree of freedom.

$$\chi^2_1 = 13.93 \quad \text{and so} \quad P < 0.001$$

Hence our assumptions of no selection, no mutation, and no migration are shown to be invalid and further study of the population is required in order to find the cause of the discrepancy.

Analysis of continuous variation

It was pointed out in Chapter 1 that there is no fixed number of classes when the variation studied is continuous. This is true whether the variation studied is measured in units of length, time, or weight on the one hand or is meristic on the other. The length of leaves, the tolerance of toxic metals in the soil, or the time of flowering are all characters which can be analysed in essentially the same way. In addition, the number of petals or the number of internodes can also be treated as continuous variables, as mean values which are not whole numbers can be sensibly calculated.

Let us take the number of flowers per plant as an example. If we count the number of flowers produced by each plant of a given species in a wild population, then there are two statistics we can readily determine. We can calculate the *mean* (the average) number of flowers per plant by

$$\bar{f} = \sum_{i=1}^{n} f_i/n \qquad (1)$$

(where f is the number of flowers on a particular plant and $i = 1$ to n is the number of plants) and we can determine the range of flower number. The range does not give any indication of the general spread of flower number, because an extreme individual may greatly accentuate the range. A more useful statistic is the *variance* which is a measure of the spread about the mean. Variance is related to the sum of the squares of the difference between the sample mean and, in this example, the number of flowers on each plant. Here:

$$V(f) = \sum_{i=1}^{n} (f_i - \bar{f})^2/(n-1) \qquad (2)$$

i.e. the variance of f equals the sum of, from 1 to n plants, the squares of the deviations from the sample mean divided by the number of degrees of freedom (see Appendix I), where n is the number of plants in the sample. This formula is not easy to use, so instead the algebraic equivalent

$$V(f) = \left[\sum_{i=1}^{n} f_i^2 - \left(\sum_{i=1}^{n} f_i\right)^2 \Big/ n\right] \Big/ (n-1) \qquad (3)$$

is preferred. All one has to do is to add up the flower numbers $\sum f_i$ and the squares of the flower number $\left(\text{i.e.} \sum_{i=1}^{n} f_i^2\right)$ and feed these into formula (3) to obtain the variance. A short-hand notation for the right-hand side of equations (2) and (3) is SS $(f)/(n-1)$ where SS stands for sum of squares.

We have designated the variance as V but there is an alternative notation more commonly used in biological statistics and that is s^2. This symbolism refers to the variance of a sample. If we know the true variance of the population (from which the sample came) we can use σ^2.

To make the distinction between s^2 for sample variance and σ^2 for population variance is being very strict, as Mather is in his books, but some authors use s^2 and σ^2 indiscriminately and it is as well to be aware of this.

There are several useful properties of variances which can be used during analysis. The square root of the variance (i.e. $\sqrt{V} = s$) is the *standard deviation*. When variances are added the sum is still a variance. When a larger variance is divided by a smaller one we perform an F, or variance ratio, test and this is used to determine whether two variances differ significantly from each other (see Appendix II). There are, however, several rules governing the addition and division of variances which will not be given here.

Analysis of variance

There is no technique of wider application to the analysis of variation in biological systems than the analysis of variance. Unfortunately it is often considered conceptually difficult and in consequence is left out of introductory books of statistical methods. Both Mather and Campbell treat this scepticism with contempt and describe the methods which in their experience are the best, taking the attitude that any analysis worth doing is worth doing properly. The danger of using the less powerful methods, introduced in books on statistical methods not biologically orientated, is the possibility of confusion rather than clarity when it comes to interpretation of experimental results. After all, we resort to statistical methods to help us to distinguish between different explanations of results and it is as well to use those which will do so with the most certainty. This is not to say that graphs, tables, pie charts, etc., are not useful for *presenting* data; the trouble comes when trying to *interpret* the data by these methods. Two worked examples showing the use of the analysis of variance will be found in Appendix II.

Genotype and continuous variation

With discontinuous variation, plants of the same phenotype generally contain the same genes to determine that phenotype. On the other hand, with continuous variation it is possible that the same phenotype may arise from different combinations of several genes. According to the genetic similarity of individuals and the breeding system of the population from which they were obtained, there are predictable results to be expected from a breeding programme.

(*a*) If two plants of the same phenotype are also of the same genotype then the progeny produced by self-fertilization should show the same mean and variance as the progeny obtained by crossing the two plants. When the parents differ in genic content then the progeny of a cross should show a greater variance than the progeny of each of the selfs.

(*b*) If the plant is an inbreeder then on selfing there should be a low variance in the phenotypes of the progeny so long as they are grown in a uniform environment. Outbreeders, when selfed, should show a large variance in their progeny. To be doubly sure, it is useful to raise a second generation obtained by selfing. If the original parent was homozygous at most loci then the second-generation families should have the same variance as the first within the limits of sampling error and variations in the environment. As a control, known inbred lines can be grown in the same experiment in order to assess the variance due to the environment. Using this breeding programme it is then relatively easy to determine the relationship between two plants with the same character.

Diagrammatic presentation of data

The most thorough way of presenting data is in tabular form, but unfortunately a table can be indigestible. Consequently various techniques have been developed for simplifying the presentation in order to improve the ease of interpretation. Although pictorial representation of data is discussed before tests of significance in most books on this topic, it has been deliberately left over until the end of this chapter. Too often the presentation of data is emphasized more strongly than the analysis and interpretation and this may well be a consequence of the order in which the information is usually presented in textbooks.

The present trend of arithmetic teaching, even in reception classes of infant schools, is towards an understanding of number and much of the time the children are constructing histograms of the distribution of foot size in the class or line graphs of the temperature in the playground at breaktime. Hence by the age at which this book becomes of interest, the reader should be well versed in, if not entirely nauseated by, data presentation.

We do not in any way belittle the value of histograms (Figure 2.1), pie charts (Figure 2.2), and scatter diagrams (Figure 2.3), for not infrequently a scatter diagram, for example, is so indicative of a correlation that statistical analysis is unnecessary (see Figure 2.3). On other occasions, the plot will suggest whether the relationship between two variables is linear or exponential and so, by plotting the data first, a clue to the correct analytical procedure may be gained. Eventually, however, there is no substitute for experience and therefore the best recommendation that can be given is that practice should be gained in as many variations of analysis of biological data as possible.

Experimental cultivation methods

We must now return to the practical business of measuring variation with particular reference to plasticity. The phenomenon of plasticity in plants (which we shall cover in detail in Chapter 3) clearly presents some severe difficulties. It is essential for the understanding of natural selection to try to separate the two

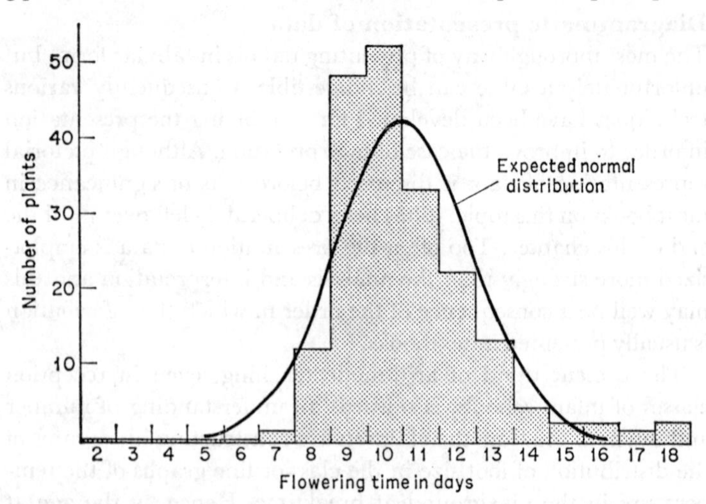

Fig. 2.1. *Arabidopsis thaliana* (thale cress). The plants included in this histogram were obtained from seed collected in one population of the plant. Flowering time was measured as the number of days after sowing of the seed that the first flower opened. The histogram approximates a normal distribution. (Modified from Jones, M.E. 1971.)

influences, those of the genotype and of the environment, which interact to produce the observed phenotype. Put like this the proposal is hopeless—the genotype and the environment are equally essential to the development of a plant and it is meaningless to ask what the plant would have looked like in the absence of an environment. What we can do is make comparisons. When two plants differ in phenotype they may do so because they are genetically different, or because they developed in different environments, or both. We run into trouble if we try and put it the other way round and ask whether or not two plants are identical—although an interesting example of this will be described in Chapter 6—the statistical ideas already discussed assume that we are all the time concerned with the significance of differences.

There are two types of experiment which can be designed. We can grow all our plants under uniform environmental conditions and assume that any differences which remain between the phenotypes have a genetic basis. On the other hand we can divide a suitable plant into pieces and grow these genetically identical individuals under different environmental conditions. Any differences of phenotype must then be due to the effects of environmental differences during development. It is the first of these experiments with which we are normally concerned when we are looking for genetic differences between population samples, and

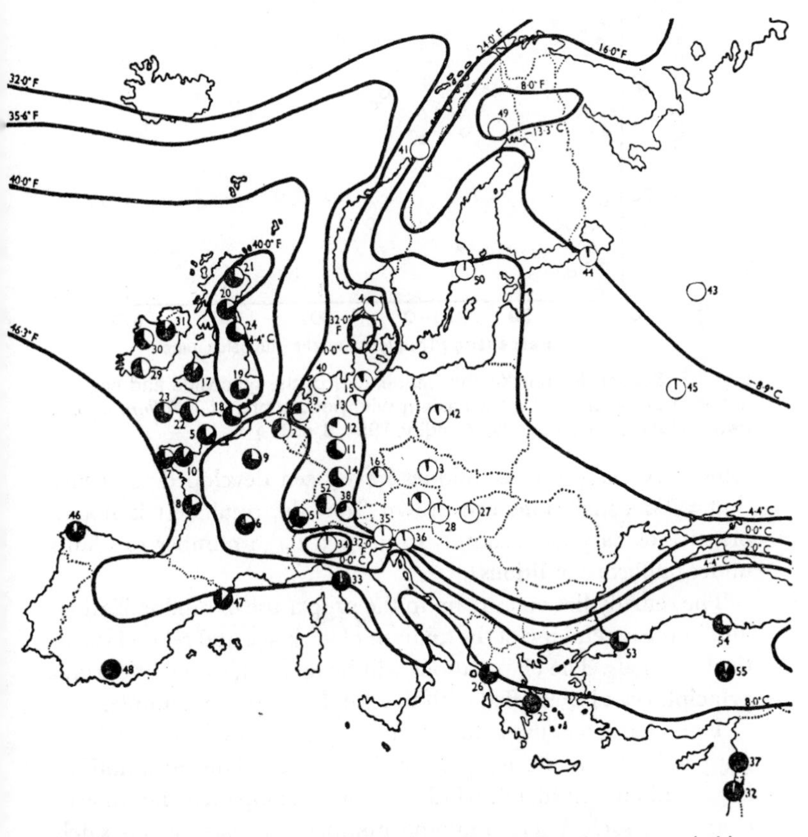

Fig. 2.2. The distribution of the cyanogenic form of *Trifolium repens* (white clover) in Europe. The black section of each circle (pie) indicates the frequency of the cyanogenic form. (Modified after Daday, 1954.)

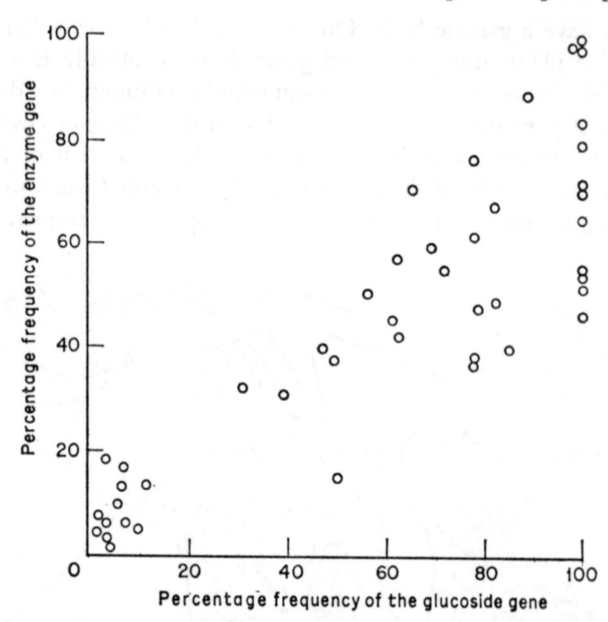

Fig. 2.3. Scatter diagram of the frequencies of the glucoside and enzyme genes in European and Near Eastern wild populations of *Trifolium repens* (white clover). (After Daday, Heredity, 1954, **8**, p. 61.)

some very elaborate techniques have been developed to ensure that valid conclusions are drawn from the results. It is not in practice as easy as it sounds to grow a large number of plants under uniform conditions.

The remedy lies once again in the proper use of randomization, and in the careful use of the analysis of variance and related statistical methods. The elaborations will be ignored, as the important principles can be clarified with the simplest possible example.

The character used to illustrate the analysis of variance (Appendix II) is the number of flowers per plant, and nothing is said about plasticity. Had it been a completely non-plastic character, such as chromosome number, the data could safely have been collected from populations growing in the wild. The danger of doing this with flower number should now be obvious.

The difference in mean flower number (22.1 per plant in population *A* compared with 20.0 in population *B*) could just as well be due to a difference of environment as to a difference of genetic constitution. A population growing in the shade, for instance, would be likely to flower less freely than one growing in the open.

We thus have to inquire under what experimental conditions the data could have been obtained so as to lead to the firm conclusion that the difference was one of genetics. The simple answer, controlling the environment, is unfortunately not a practical one. No matter how carefully the plants are grown in controlled environment chambers it is never possible to say that there is absolutely no difference between the conditions experienced by separate plants. The proper answer is to spread out the effects of the inevitable environmental differences between all the individuals of the two samples. This requires no elaborate equipment, merely a reasonably uniform field in which the plants can be grown in a strictly randomized layout so that the *average* environment for one sample is exactly the same as that for the other. We can then count the flowers and carry out the analysis which compares the two sample means. A significant difference between them *suggests* that there is a real difference of genetic constitution affecting the number of flowers per plant. The heritable nature of flower number and the details of the genetic mechanism can only be confirmed by an appropriate breeding programme.

Much of the early work on morphological variation was done in the experimental garden, and once techniques like the analysis of variance became available remarkably precise and detailed work could be done with a minimum of equipment. The picture is rather different with some of the physiological studies discussed in Chapter 4, when the use of controlled cultivation facilities and elaborate experimental techniques is inevitable. In spite of this the same basic principles of experimental design have to be observed. It is still essential to distinguish between genetically and environmentally controlled differences. The same problems of randomization and replication have to be solved if we are to draw valid deductions from even the most sophisticated experiments.

This is sometimes far from easy. Whereas hundreds, or even thousands, of plants can be grown in a field and measured with ease it is no simple matter to put an equally large sample through an exacting and intricate series of physiological treatments. It has to be admitted that even some of the most convincing and carefully executed work of this kind can to some degree be faulted because of its concentration on a small number of individual plants.

There are several important principles which must be considered when using experimental cultivation techniques for the study of continuous variation. If plants are to be grown from seed then all the seeds for one experiment should be sown in sterile seedling compost on the same day. All seedlings should, as far as is possible, receive the same treatment, and all the established plants should be put into the experimental field on the same day.

It is also essential to have a high germination rate and this is particularly important when the seeds normally have a dormancy phase. If germination is low we have no means of determining whether we have a random sample of the genotypes in the population among the survivors, since there is the danger that we will select those genotypes which give rise to a high germination rate. If the latter is the case any results obtained with the mature plants may be meaningless.

The actual arrangement of the plants in the field generally follows a randomized block design and it is usual to have a marginal group of plants, which are not part of the experiment, round the perimeter of the block to act as 'guards'. They ensure that the experimental plants at the edge of the plot are subjected to the same competition from neighbouring plants as are the others in the body of the plot.

Where the character under study shows discontinuous variation careful planning is not so important but it is useful to adopt the same procedure merely for the sake of consistency and hence of simplicity.

The problems discussed in this section have been treated only in outline. Books on biological statistics should be consulted for a fuller, and thereby more satisfactory, treatment.

3

Effects of the environment on variation

Having discussed the methods used for sampling populations of plants and for analysing the data, we must now extend consideration of the origin and types of variation to the processes by which changes in the frequency of the variant forms may arise. Three basic types of selection can be distinguished and we will consider these in relation to continuous variation.

Types of selection

In Figure 2.1 we showed the frequency distribution of flowering time in a particular population of *Arabidopsis thaliana* (thale cress) and we superimposed upon it the normal curve with the same mean and variance as the basic data. Experience shows that a normal distribution is common to a great many characters which show continuous variation and, indeed, most of the statistical techniques used in the analysis of both biological and non-biological data assume the data to be distributed normally. On occasion we actually change the scale, for example, by taking logarithms and making the data normal so that they can be analysed.

With biological characters, the variance of the distribution can sometimes provide more than just a description of the spread of the variation. Suppose there is independent evidence that selection favours the average phenotype at the expense of the extremes. The mean will be essentially the same as the optimum and extreme forms will become rare. The effect of such **stabilizing** is to reduce the variance. A change in the intensity of stabilizing selection will result in a change in the shape of the distribution curve and this is shown in Figure 3.1. We can therefore use the

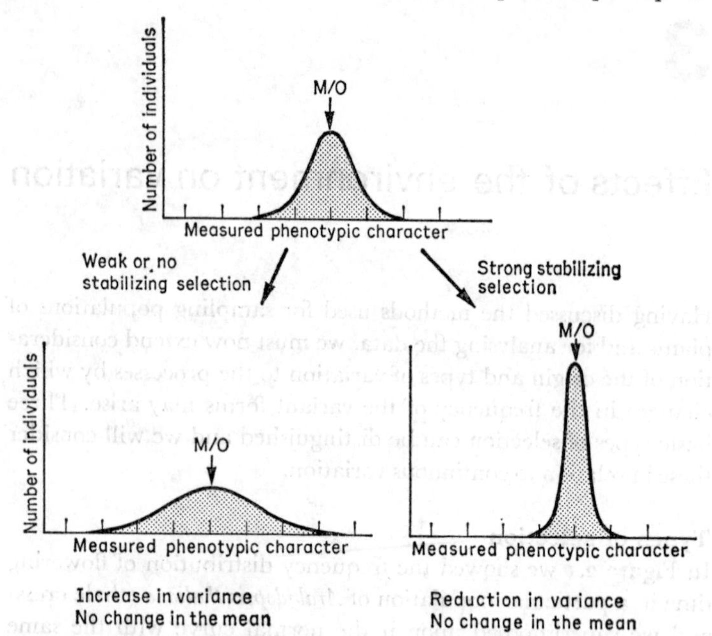

Fig. 3.1. The effect of stabilizing selection (M indicates the mean phenotype and O indicates the optimum phenotype).

shape of the curve as a guide towards the next stage in the investigation of the intensity and the agencies of selection.

The most frequently discussed example of stabilizing selection is concerned with wing span in British sparrows (*Passer domesticus*) in North America. In 1898, Bumpus measured the wing span of 176 sparrows killed during a storm and found that, compared with the survivors, there was an excess of birds with wings shorter or longer than the average. With plants it is more difficult to find examples but it is likely that flowering time in *Papaver dubium* (long-headed poppy) and *Nicotiana rustica* is under stabilizing selection.

Rather than having a situation in which selection favours the mean we may find that selection favours one of the extreme phenotypes more than the other. Here the optimum phenotype is not the same as the mean. In response the mean phenotype of the population shifts towards the optimum, because the unfavoured forms die earlier or produce fewer progeny than their

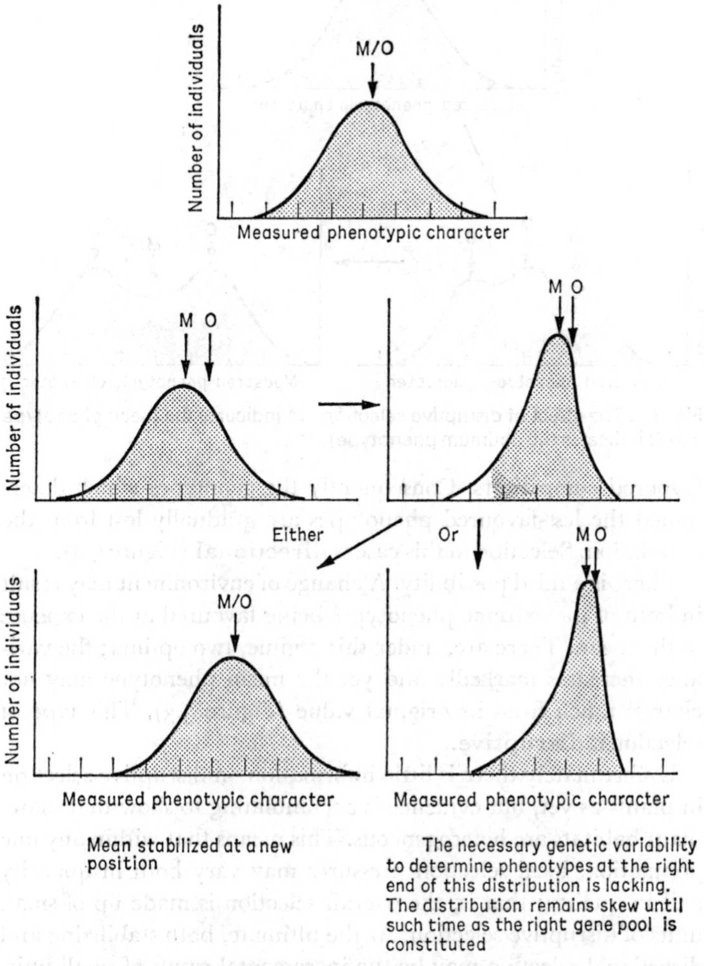

Fig. 3.2. The effect of directional selection (M indicates the mean phenotype and O indicates the optimum phenotype).

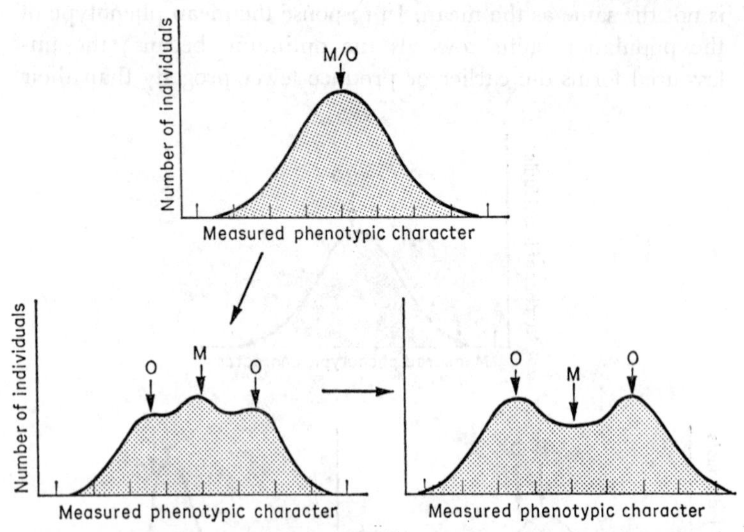

Fig. 3.3. The effect of disruptive selection (M indicates the mean phenotype and O indicates the optimum phenotype).

favoured competitors. Consequently the genotypes which determined the less-favoured phenotypes are gradually lost from the population. Selection in this case is **directional** (Figure 3.2).

There is a third possibility. A change of environment may result in both of the extreme phenotypes being favoured at the expense of the mean. There are, under this regime, two optima; the variance increases markedly and yet the mean phenotype may not change much from its original value (Figure 3.3). This type of selection is **disruptive.**

Unfortunately there is little information on disruptive selection in plants as yet, but evidence is accumulating to show that many plant habitats are heterogeneous. This means that within any one population, local selection pressures may vary both in quantity and in quality; that is, the overall selection is made up of small units of disruptive selection. At the ultimate, both stabilizing and directional selection may be the incremental result of small units of differential disruptive selection.

The highly complex nature of the environment in which a plant lives has been appreciated for a very long time, yet it is only during the last few years that techniques of analysis, let alone working hypotheses, have been developed which allow us to begin to break down the various components of the environment in ways which make sense to us and to our instruments. Whether these same categories mean much in terms of plant biology is another matter and we shall touch on some aspects of this problem later.

Changes in gene frequency by selective processes

Before examining the processes by which changes in gene frequency occur, we must first tackle the relationship between the allele frequency in one generation and the next. We will do this by developing a model. Consider a pair of alleles A_1 and A_2 in a population of an annual plant species where the frequencies are p and q respectively, given that $p + q = 1$. There are three possible genotypes A_1A_1, and A_1A_2, and A_2A_2 and if we suppose that individuals of each of the three genotypes produce the same number of fertile gametes, then the relative frequencies of the two gametic types will be the same as the allele frequencies. We have an array of p A_1 gametes and q A_2 gametes. If, further, we assume that gametes can unite at random, then the genotypes of the progeny and their frequencies will be as shown in Table 3.1.

Table 3.1

The frequency of genotypes in a random mating population.

		Male Gametes	
		pA_1	qA_2
Female Gametes	pA_1	$p^2A_1A_1$	pqA_1A_2
	qA_2	pqA_1A_2	$q^2A_2A_2$

From the table we can extract the frequencies of the three possible phenotypes:

$$\begin{array}{ccc} A_1A_1 & A_1A_2 & A_2A_2 \\ p^2 & 2pq & q^2 \end{array}$$

We know that

$$p^2 + 2pq + q^2 = (p + q)^2$$

and that

$$p + q = 1$$

so

$$(p + q)^2 = 1.$$

Let us assume that the population is large and now carry on to the next generation. With three genotypes there are nine possible types of mating and the frequency of each of these is the product of the frequencies of the appropriate genotypes. The matings and their progeny are listed in Table 3.2.

Table 3.2

The nine possible matings between individuals of three genotypes and the frequency of the resultant progeny.

Mating			Progeny		
Female		Male	A_1A_1	A_1A_2	A_2A_2
$p^2A_1A_1$	×	$p^2A_1A_1$	p^4	—	—
$p^2A_1A_1$	×	$2pqA_1A_2$	p^3q	p^3q	—
$2pqA_1A_2$	×	$p^2A_1A_1$	p^3q	p^3q	—
$p^2A_1A_1$	×	$q^2A_2A_2$	—	p^2q^2	—
$q^2A_2A_2$	×	$p^2A_1A_1$	—	p^2q^2	—
$2pqA_1A_2$	×	$2pqA_1A_2$	p^2q^2	$2p^2q^2$	p^2q^2
$2pqA_1A_2$	×	$q^2A_2A_2$	—	pq^3	pq^3
$q^2A_2A_2$	×	$2pqA_1A_2$	—	pq^3	pq^3
$q^2A_2A_2$	×	$q^2A_2A_2$	—	—	q^4
		Totals	$p^4+2p^3q+p^2q^2$	$2p^3q+4p^2q^2+2pq^3$	$q^4+2pq^3+p^2q^2$

The progeny totals can be rearranged as A_1A_1, $p^2(p^2 + 2pq + q^2)$; A_1A_2, $2pq(p^2 + 2pq + q^2)$; and A_2A_2, $q^2(p^2 + 2pq + q^2)$ which because $p^2 + 2pq + q^2 = 1$ reduce to p^2, $2pq$, and q^2

respectively. These frequencies are exactly the same as in the previous generation. The generalization, that in a random mating population the allele frequencies remain constant from generation to generation, was suggested simultaneously, but independently, by G. H. Hardy and W. Weinberg in 1908, and is now called the Hardy–Weinberg law.

But there are certain inbuilt assumptions. We have not allowed for selection, migration or for mutation.

Models for selection

What effects can selection have? It may be that there is a differential survival rate between the various phenotypes up to the time of full maturity. Alternatively, or in addition, the phenotypes may

Table 3.3

Modification of the Hardy-Weinberg model to allow for selection and dominance. Note that if $d = 0$, A_1 is dominant; if $d = 1$, A_2 is dominant.

If a certain proportion s of A_2A_2 individuals do not survive to reproduce, then $1 - s$ A_2A_2 individuals do survive. s is therefore a measure of the selection against the phenotype determined by A_2A_2: s is the selective coefficient.

	A_1A_1	A_1A_2	A_2A_2	Total
Frequency	p^2	$2pq$	q^2	1
Actual numbers in a population of size N	Np^2	$2Npq$	Nq^2	N
Relative selective value if A_1 is dominant over A_2	1	1	$1 - s$	
Allowing for heterozygote being of intermediate phenotype	1	$1 - ds$	$1 - s$	

vary in their production of gametes. Fortunately, it is possible to calculate the selective advantage or disadvantage of a phenotype independently of the stage of the life-cycle at which selection acts for the contribution to the next generation is measurable entirely by the total gametic output of all the individuals of a given phenotype. Leaving aside mutation and migration for the moment, let us consider the effect of selection and dominance on the Hardy–Weinberg model. Suppose that selection is acting against the phenotype determined by the homozygote A_2A_2 where s is a measure of this selection. Table 3.3 can be constructed.

It matters little at this stage whether we consider that some A_2A_2 individuals are removed from the population or fail to reproduce, or that all the A_2A_2 survive but each produces a smaller number of gametes than A_1A_1 and A_1A_2 individuals for the net effect is the same. For simplicity of the model, however, it is useful to consider that some A_2A_2 fail to reproduce or die. The frequency of the three genotypes after selection is shown in Table 3.4.

Table 3.4

The frequency of the three genotypes after selection. Note that $N - N(1 - sq^2 - 2dspq)$ individuals have been lost from the population.

A_1A_1	A_1A_2	A_2A_2	Total
Np^2	$2Npq(1-ds)$	$Nq^2(1-s)$	$N(1-sq^2-2dspq)$ *

$$^*Np^2 + 2Npq(1 - ds) + Nq^2(1 - s) = N(p^2 + 2pq - 2dspq + q^2 - sq^2)$$
$$= N(1 - sq^2 - 2dspq) \text{ because } p^2 + 2pq + q^2 = 1.$$

The frequency of the A_1 alleles before selection in the next generation will be:

$$p = \frac{2Np^2 + 2Npq\,(1 - ds)}{2N\,(1 - sq^2 - 2dspq)} = \frac{p^2 + pq - dspq}{1 - sq^2 - 2dspq} \qquad (1)$$

which is independent of the size N of the population. The change in the frequency of A_1 is written Δp and is:

$$\begin{aligned}\Delta p &= p_1 - p_0\\ &= \frac{p^2 + pq - dspq}{1 - sq^2 - 2dspq} - p\\ &= \frac{p(p + q - dsq) - p(1 - sq^2 - 2dspq)}{1 - sq^2 - 2dspq}\\ &= \frac{p(1 - dsq - 1 + sq^2 + 2dspq)}{1 - sq^2 - 2dspq}\\ &= \frac{spq(q + 2dp - d)}{1 - sq^2 - 2dspq} \qquad (2)\end{aligned}$$

where p_0 is the original parental frequency of p(i.e. $p_0 = p$ in this example). Notice that the Δ p we have calculated is related only to the value of p, *s*, and *d* of the previous generation. Hence if we know p, Δ p, and *d* we can calculate *s*. For simplicity the dominance parameter *d* will be taken as 1 or 0 for the moment. Which allele is dominant can be determined by the appropriate breeding programme. When A_2 is recessive, $d = 0$ and

$$\Delta p = spq^2/(1 - sq^2) \qquad (3)$$

In the models we have been considering up to now we have been assuming random mating and that the plants are annuals. We can build into the models parameters allowing for inbreeding and other forms of mating system as well as perenniality and other types of overlapping generations. All these variations of the breeding system and longevity convey selective advantage or disadvantage on those individuals showing particular phenotypes and hence would be included in the selective coefficients built into the models. Which of the various selective agents is acting must be determined by studying the general ecology of the plant.

Furthermore, we have been assuming that the populations are discrete, that is, the distribution of the plants is known and that the margin of the populations are clear-cut. This is, however, a naïve approach because the majority of species are distributed in a continuous manner over a very wide area. The consequences of this will be discussed later in this chapter.

We will now move on to consider, from a theoretical point of view, the two classes of discontinuous variation discussed in the first chapter in relation to the models we have developed in this one.

(1) If we have a situation where a common dominant allele A_1 is mutating to the rare A_2 faster than A_2 reverts back to A_1, there will be a net increase in the frequency of A_2 which we will call μ p (where μ is the net mutation rate per generation and the actual number of mutations is proportional to the value of p). We noticed earlier in this chapter (equation 3) that when A_1 is dominant over A_2

$$\Delta \mathrm{p} = s\mathrm{pq}^2/(1 - s\mathrm{q}^2)$$

But when q is small (A_2 being rare) q^2 is very small and so the denominator is approximately equal to 1. Under these circumstances

$$\mathrm{p} = s\mathrm{pq}^2.$$

At equilibrium, the increase in p due to selection is exactly balanced by the decrease due to mutation, and hence

$$\Delta s\mathrm{pq}^2 = \mu \mathrm{p} \text{ or } \mu = s\mathrm{q}^2 \qquad (4)$$

Thus if we know s or μ we can find the other in terms of q.

Referring back to Table 3.4, the actual number of individuals which fail to reproduce is $\mathrm{N} - \mathrm{N}(1 - s\mathrm{q}^2 - 2ds\mathrm{pq})$ which equals $\mathrm{N}s\mathrm{q}^2$ when $d = 0$ (i.e. A_1 is dominant over A_2). The proportion which fails to reproduce is $\mathrm{N}s\mathrm{q}^2/\mathrm{N}$ which reduces to $s\mathrm{q}^2$; but we have already shown that $\mu = s\mathrm{q}^2$, so we have to conclude that an equilibrium can be reached when the number of individuals which die or fail to reproduce for other reasons equals the mutation rate.

(2) What are the genetical models which will explain a polymorphism? You will remember we argued that a polymorphism was maintained by a balance of selective forces. One of the simplest models is the two-allele case when both homozygotes are

at a disadvantage with respect to the heterozygote. We cannot assume that selection against A_1A_1 and A_2A_2 individuals is the same so consider the model in Table 3.5.

Table 3.5

Model for selection against both homozygotes.

	A_1A_1	A_1A_2	A_2A_2	Total
Genotype frequencies	p^2	2pq	q^2	1
Actual numbers in a population of size N	Np^2	2Npq	Nq^2	N
Relative selective values	$1 - s_1$	1	$1 - s_2$	
Actual numbers of those which reproduce	$Np^2(1 - s_1)$	2Npq	$Nq^2(1 - s_2)$	$N(1 - s_1p^2 - s_2q^2)$ *

*$N[p^2(1 - s_1) + 2pq + q^2(1 - s_2)] = N(p^2 - s_1p_2 + 2pq + q^2 - s_2q^2) = N(1 - s_1p^2 - s_2q^2)$

$$\Delta p = \frac{2Np^2(1 - s_1) + 2Npq}{2N(1 - s_1p^2 - s_2q^2)} - p$$

$$\Delta p = [p^2(1 - s_1) + pq - p(1 - s_1p^2 - s_2q^2)]/(1 - s_1p^2 - s_2q^2)$$

$$\Delta p = pq(s_2q - s_1p)/(1 - s_1p^2 - s_2q^2)^{**} \qquad (5)$$

At equilibrium, when the selective forces are balanced, $\Delta p = 0$; that is, the frequency of the A_1 allele remains constant from

** $p^2(1 - s_1) + pq - p(1 - s_1p^2 - s_2q^2) = p^2 - s_1p^2 + pq - p + s_1p^3 + s_2pq^2 = pq(s_2q - s_1p)$

because $p^2 - p = p(1 - q) - p = p - pq - p = -pq$

and $s_1p^3 - s_1p^2 = s_1p^2(1 - q) - s_1p^2 = s_1p^2 - s_1p^2q - s_1p^2 = -s_1p^2q$

generation to generation. This will be satisfied in equation 5 when $p \times q = 0$ or when $s_1p = s_2q$. The former will be the situation if either allele becomes fixed, while the latter will be true for a balanced polymorphism. Rearranging the equation it becomes $p/q = s_2/s_1$ or $p = s_2/(s_1 + s_2)$.

The equation 5 can be used to demonstrate how the balance is maintained. If p increases a little, e.g. by migration from another population with a different allele frequency s_1p will be larger than s_2q and so Δp will be negative. The effect is, therefore, to reduce the value of p. When q increases, on the other hand, s_2p is larger than s_1p and so Δp will be positive. Hence p increases. Note, however, that p must not become either 1 or 0, for if this does occur pq would be zero and one or other allele would be eliminated from the population.

Comment

This type of model building may appear to bear little relation to what actually happens in wild populations of plants and animals and in detail this is true, but a most useful overall picture can be gained by these means. When we began with the Hardy–Weinberg law we said that there were certain inbuilt assumptions about the absence of selection, mutation, and migration. Significant departures from the Hardy–Weinberg law (see the example in the previous chapter) are usually due to selection acting differentially on the various phenotypes and thus by using the test it can be possible to detect selection. But only further examination of the population will, perhaps, reveal the type—viz. directional, stabilizing, or disruptive—and its intensity and the actual agents involved.

Changes of gene frequency by random processes

In the first section of this chapter we discussed the three types of selection which can be recognized. There is little doubt that directional selection has a powerful influence on the frequency of alleles in a population, but major difficulties arise when one comes to consider the effects of random processes. If a small number of organisms are withdrawn from a population it is unlikely that they will contain a representative sample of all the various genes present in the original population. When a new population is founded in a new area, or a depopulated area, there is only a limited number of genes available and so the genetic make-up of the immigrants will predetermine the variation available in the new population and hence the differences between the new and the original population may well be due merely to the effects of random sampling. This founder principle was first clarified by Mayr (1954) and no one disputes the logic and value of his argument.

A controversy rages, however, over whether errors of random sampling are important within existing populations. Wright and many other North American authors, maintain that random fluctuations in gene frequency (genetic drift) are important, but Fisher and Ford (see Ford, 1971, for references) have suggested that genetic drift is so trivial in relation to selection that it can generally be ignored. Both sides agree that it can only be of consequence in small populations, but Fisher and Ford have pointed out that those selection pressures so far measured in wild populations of plants and animals are so large that any effect of random fluctuations would be completely swamped.

Fisher and Ford have, perhaps, overstated their case for even though all attempts at demonstrating it by experimental means, let alone by observation, have failed, it is clear from population models that genetic drift must occur. Why, therefore, should drift be so difficult to demonstrate? It is essentially because the process is a random one, taking place in the absence of or in spite of selection, mutation, and migration and it is very difficult to *show* that none of these factors is acting however easy it is to *imagine* it. Migration can be controlled and allowances can be made for mutation but it is an entirely different matter to reduce selection to zero, essentially because in no case are all the selective agents known.

To demonstrate genetic drift it is first necessary to find alleles which are selectively neutral with respect to each other. It has been suggested that it may be possible to find enzymes with identical physical and chemical properties which differ in one or more amino acids. The genes which determine these enzymes would differ in only a small number of DNA base pairs. If enzymes with these properties existed then the experiment would involve starting with populations of the organism of different size and of different allele frequency, exerting strictly controlled environmental conditions, and seeing what happened to the allele frequencies. Although elegant in conception, such an experiment is full of the most dreadful pitfalls, and interpretation of the results is hazardous.

Are there any characters upon which selection has acted in the past but which now confer no selective advantage or disadvantage on the individuals possessing them? From what is known of discontinuous variation we would fail to find examples of neutral characters here. Thus only among characters showing continuous variation could such a situation be found.

Consider the number of petals in *Ranunculus ficaria*, where the range is from 7 to 12 with different numbers being found in flowers on the same plant. In other parts of the genus the petal number is more regularly five, see Table 3.6. Why should this be?

Table 3.6

Petal number in the genus *Ranunculus.*

Subgenus	Petal number
Ranunculus	5 (sometimes less, i.e. *R. sardons* and *R. parviflorus*)
Batrachium	5 (5 – 10, *R. fluitans*)
Ficaria	0, 7 – 12

Sepals, petals, and stamens are modified leaves and in those species in which petal number is variable petaloid stamens are not infrequently seen. A unit which is part stamen and part petal is presumably the result of some error of development. Which developmental process, petal or stamen, started first cannot be determined with confidence merely by observing the open flower, but it is apparent that neither process can be rigidly controlled.

It can be argued, that where petal number is relatively constant there is strong selection in favour of stability in the development of the character. Quite why is another question. In other closely related species, however, the selective process has not been strong enough to limit in any strict fashion the number of petals which develop. It is possible that, in those species where the petal number varies considerably from flower to flower on the same plant, the coefficient of selection is very low and may even approach zero. Therefore, such a species may be suitable for experiments on genetic drift. Even if a plant with suitable variation of petal number is found, the usual difficulties over the breeding system and other aspects of the general biology of the species are bound to create experimental problems which will have to be overcome first. Another discussion on the variable number of parts in the flowers of *Ranunculus ficaria* can be found in Briggs and Walters (1969).

Clines and ecotypes

The classifying habit of the mind is a very valuable scientific instinct, but there are some situations in which it is inappropriate. Taxonomy naturally emphasizes characters which show clear discontinuities, such as leaves which are simple or compound, hairs which are present or absent, or petals which number three, four, or five. Most species seem to be separated from each other by reasonably clear gaps. Below the species level, however, much of the variation we study is not suitable for this sort of treatment, and as a result a certain amount of special terminology has had to be invented to handle the infraspecific mixture of continuous and discontinuous variation.

The term 'cline' was proposed by Huxley (1942) to signify a character gradient—in other words a gradual change in a character as you travel through the geographical range of a species, rather than an abrupt change at a certain point where one form gives way to another. It is important to emphasize that this can occur with any kind of character. Even a character which varies discontinuously from plant to plant may still show clinal variation on a wider scale. Indeed the term was first applied to a series of populations in which the proportions of two distinct morphological forms of Guillemot changed from south to north up the coast of Britain. A good example in plants also involves a north–south gradient. *Arum maculatum* (wild arum) may have leaves which are spotted or plain, and very many populations show a mixture of the two, i.e. they are polymorphic. When the relative proportions of the two forms are considered the spotted morph is found to decline in frequency as we go northward (Prime, 1960). The two forms are distinguished by a single major gene and the gradient is best thought of as a change of allele frequency.

Clines are even more frequent in characters showing truly continuous variation. Böcher has demonstrated gradients of leaf size and shape across Europe in a number of species, including *Veronica officinalis* (common speedwell) and *Prunella vulgaris* (self heal). Here again the cline can be thought of as a gradient of gene frequency, but this would be difficult to demonstrate in the case of characters controlled by large numbers of genes. In many such instances the genetic situation has not been adequately studied, and a cline can then only be described in terms of phenotypic frequency.

The cline terminology has one disadvantage: it tends to imply that a line can be drawn on a map in such a way as to summarize all the important variation in the character under discussion. There are several situations in which this is not true. When a large amount of information is available about the variation in a character over a wide area we can plot the gene frequency at each sample point on a map, join up the points with similar frequencies, and identify areas of high and low frequency. This has been done extensively for human blood groups, and for one or two characters in wild plants. The ability of *Trifolium repens* (white clover) to produce hydrogen cyanide when damaged will be discussed in Chapter 4, but the European distribution of one of the phenotypes involved has already been shown in Chapter 2 (Figure 2.2) It will be seen that it would be possible to talk about an east–west cline, but the details are far too complex for this to be regarded as a sufficiently comprehensive description.

Before this discussion can be taken any further we must examine more closely the idea of a gradient on a map. So far we have spoken as if mere distance from point to point was the thing that mattered. In reality, of course, the environment changes as we go north in Britain, or west across Europe, and there is no doubt that it is these changes which are influencing the plants and not distance as such. If that is so we can begin to ask whether the observed changes of allele frequency are in any sense related to the changes in the environment. Are the spotted *Arum* leaves better suited to conditions in the south of England, and the unspotted ones to the climate of Scotland? As with most questions of this type we do not know the answer, but we shall discuss some of the better understood examples in Chapter 4.

This idea that clinal variation might reflect adaptation to environmental gradients allows us to make an important extension to the whole cline concept. There is no longer any need for a cline to show large-scale trends on a map at all. Once we have said that some environmental factor, and not distance, is responsible for selection we are in effect looking for a correlation between allele frequency and the appropriate habitat factor. For instance, to anticipate another example from the next chapter, in areas of Britain where lead has been mined there are large numbers of waste tips scattered about which may contain enough lead to be highly poisonous to plants. Other tips contain less, depending on the geology and on the efficiency of the lead extraction process. The surrounding soil usually contains virtually no lead at all.

Thus the lead content of the soil varies markedly from point to point, but all values from high to low can be found by sampling a range of tips. A number of plant species have evolved sufficient tolerance of lead to grow on all but the most poisonous tips, and in some species at least tolerance is a continuously varying character. This results in a correlation between the degree of tolerance of the plant and the concentration of lead in the soil of its particular tip (Figure 4.2).

Here we have a correlation between gene frequency and habitat which cannot be expressed as a line on a map simply because the selective factor is distributed in a complex way. We can collect data from many tips and arrange the figures in order of increasing lead content. We can then show that tolerance increases in the same order. We have built up a 'subjective cline' which is just as real as those produced in response to major climatic factors across Europe—it is merely less obvious. As an additional aid to clarity of expression any cline which has been shown to have an ecological interpretation is called an ecocline. Other prefixes have been suggested in other situations, but this is the most useful one in the context of this book.

Does this stress on ecoclines mean that habitat-related variation can never be classified? This is still a matter for debate. The term ecotype is in wide circulation as an infra-specific category. As described in Heslop-Harrison (1953) it began as a term for a morphologically distinct population typical of a particular type of habitat. It has gradually been extended in its use to cover smaller degrees of difference on the one hand, and physiological characters on the other, and it is at these lower and more cryptic levels that it is in course of being replaced by the more flexible ideas involving clines. The crucial difference between the two sets of terms is that a cline is potentially continuous, while ecotypes are marked off from each other by definite discontinuities. It has proved extraordinarily difficult to place most examples in one category or the other, usually because of lack of evidence. It is fair to say that in most cases where clines have been looked for they have been found. In very many cases the sampling has been so discontinuous that it is impossible to say what would be found if intermediate habitats were investigated.

It is important to distinguish the cases where variation is discontinuous because the habitat is discontinuous and may be thought of as potential clines, from those in which a gap seems to be found in the variation even though the habitat appears reasonable continuous. The case of *Festuca ovina* (sheep's fescue) from acid and calcareous soils to be discussed in Chapter 4 is a good example of possibly discontinuous variation due to a gap in the available habitat. There is little doubt that this is a genuinely clinal situation for all that.

To find examples of really clear-cut ecotypes we have to turn to work at a higher taxonomic level where many characters are involved. The clinal gradients in *F. ovina* and in *Arum maculatum* involve few characters (in the latter case only one) and presumably a correspondingly small number of genes. One of the most famous pieces of work exemplifying the ecotype idea was that of Clausen (1940) on *Potentilla glandulosa.* He was dealing with a wide-ranging species which showed a great deal of variation in almost all its characters. He studied a series of populations ranging in altitude from sea level to high up on the Rockies by means of experimental garden techniques. In spite of the continuous range of climates involved Clausen found fair evidence of discontinuity, with his samples falling into three or four distinct groups, and each of these was found over a different part of the altitude range. The forms were so distinct that sub-specific names had to be given to them, and although the differences included a number of characters which were of direct adaptive advantage to the plant in its native climate, they also involved many other characters whose significance was more difficult to understand.

This question of scale will be discussed more fully later. It is enough at this point to draw attention to the great variety of situations which may be covered by the term 'ecotype'. It was used by Gregor (1938) to refer to populations of *Plantago maritima* (sea plantain) differing only in their erect or prostrate growth habit, and by Clausen to refer to morphological sub-species. When we ask how distinct the various ecotypes in a species are we usually find those at Gregor's level merge into a cline when studied in more detail, while Clausen's are separated by real and important discontinuities.

Establishment and competition

Plants do not live for ever, and so if the species is to persist they have to reproduce. Some creeping perennials may go on for a very long time, but in most communities the individuals of each species are being continually replaced. Even creeping perennials regularly produce new shoots to replace the older and less active ones, so that although there may be little seed formed there is still active reproduction by vegetative means, and in the case of grasses the older parts do not survive for more than a few years. The typical life-cycle of most higher plants does involve the regular production of seed. This means that a successful species has to be able to survive through a whole sequence of stages of development. For a seed to be called viable all it has to do is to germinate. For a species to be viable in a natural community its seeds have to germinate, and then its seedlings have to establish themselves and develop a root system which is adequate to support the shoot. The two parts of the plant have to grow in a mutually balanced way to maturity, when flowers have to be formed, pollination and fertilization accomplished, and the seeds ripened and dispersed. Failure at any one of these stages may mean that the species disappears from the community.

When we discuss the adaptive differences between organisms, and try to explain why one succeeds where another fails, we must always take the whole life-cycle into account. A comparison which is restricted to the adult phase may overlook factors of decisive importance.

The situation is sometimes even more complicated. There are some cases reported where the particular requirements of the seedling stage seem to be trivial, but effectively restrict the occurrence of the whole species. The seedlings of *Alnus glutinosa* (alder) require a wet but not waterlogged soil, and so are restricted to streamsides and some types of marsh. This requirement declines after 20 or 30 days. Young saplings can be transplanted into a wide variety of soils, and the adult tree grows best where the water-table regularly falls well below the surface during the summer (McVean, 1953). It would clearly be quite impossible to predict the ecological range of this species just from experiments confined to the adult plant.

The requirements for germination itself are extremely varied, and some of these clearly equip a plant to flourish in particular habitats. The seed coats of some desert species, for instance, contain a germination inhibitor which is effective until washed away by a thorough soaking in water. This prevents germination until a substantial fall of rain has adequately wetted the soil. Without the inhibitor the seed might germinate after only a light shower, and in a climate where rain is sporadic and the showers often separated by prolonged spells of complete drought this would be fatal. If germination is only possible after really heavy rain the seedling has a good chance of producing a deep tap-root before all the water evaporates. The problem is to tide over the period between abandoning the safety of the dormant seed and establishing a root system deep enough to draw on more reliable sources of water.

Many plants have extremely subtle mechanisms which ensure the synchronization of their developmental stages with changes in the external environment. Other examples will be discussed under the heading of adaptation to climate in Chapter 4.

It has already been pointed out that wild seedlings live a very hazardous life. They usually have to grow in the immediate presence of micro-organisms, animals, and other plants. They are thus subjected both to direct attack, from fungi, insects, and grazing animals, and to competition from other plants for the resources needed for growth.

Any of these resources may be scarce, from mere space to light, water, or mineral nutrients. The intensity of the competition depends on how serious the scarcity is and on how similar adjacent plants are in their requirements. This turns out to be an unexpectedly complex topic. We can say superficially that all green plants have the same requirements; for water, light, carbon dioxide, and essential minerals. It is true that growth is impossible in the complete absence of any of these, but in practice the plant has to contend not with complete absence but with relative scarcity. The effects of competition can be studied most easily in pot experiments with plants grown from seed. It requires elaborate facilities to control the supply of minerals in such an experiment, but water is readily manipulated, and in fact many house plants regularly grow under conditions of more or less chronic water shortage. It is easily shown in such a case that the size of the plant obtained depends on the number of plants in the pot. By analogy with the crop density effects mentioned in Chapter 1, it is clear that more plants per unit of soil mean less water for each plant, and if this is the limiting factor it will lead to smaller plants. This response to reduced circumstances by a lower rate of growth is characteristic of plants in general, and is an aspect of their great morphological plasticity, which is discussed in detail elsewhere. This is in marked contrast with higher animals. If the food supply of a mammal is drastically curtailed it will die, being unable to grow much more slowly than normal. We say that animals show a mortality response to severe competition, whereas plants often show a plastic response.

Almost all real situations are much more complicated than the 'two-plants-in-one-pot' experiment, and the differences are instructive. The pot is a relatively homogeneous unit. The roots of any plant you put into it explore the whole soil volume, and the roots of any two plants become intimately mixed up. We usually keep the pot in a glasshouse under controlled conditions of temperature and water supply, and we may go so far as to provide artificial light in the winter. In practice we rarely put more than two species into one pot even for the experimental study of competition. In all these respects the wild habitat is much more heterogeneous. The soil is deep and stratified, with plenty of opportunity for deep-rooted and shallow-rooted plants to occupy quite separate levels. The seasonal cycle of temperature and daylength allows species to live to some extent in shifts, so that some are active in the spring, while others delay their period of rapid growth until the early ones have died down. The wide range of species found in any one community is connected with these facts. A complex habitat is said to have many niches, in each of which a species can grow without coming into direct conflict with other species at every stage in its life. This goes further. Each additional species, particularly if it is made up of large individuals, alters the habitat available for others. It often makes the habitat more complex than it was before, by providing new niches. This is obvious when we consider the opportunities afforded to parasites by the appearance of a new host tree in a community. There are many less obvious effects. If a wood consists of many tree species there will be great variations in the light intensity and in the type and amount of litter on the ground, and these factors will increase the variety of herbaceous plants which are found below the trees, and of course greatly increase the number of animal species which the wood can support.

This concept of niches plays a very important part in ecological theory. Its importance lies in the fact that niche formation reduces the intensity of competition. If two plants root at the same depth they are in direct competition for food and water, while if they root at different depths that competition is much reduced. The prediction from this is that the more similar two individuals are the more intense is the competition between them. There are one or two experiments which illustrate this very convincingly.

Harper's experiments (1961) with four British species of *Papaver* (poppies) were designed to show the effect of increasing the density of plants, and of replacing plants of one species with those of another to make various mixtures. He started with seed, and sowed each plot with a different density of two species in carefully planned proportions. The densities used were so high that even the great plasticity of which poppies are capable was not sufficient to allow them all to survive. There was just not enough of the essentials of life to go round, and the result was a high degree of differential mortality. The most informative results from this experiment are those which reflect this mortality, by comparing the number of mature plants surviving at the end with the number of seeds sown at the outset. This can be thought of as the success rate.

Each part of the experiment involved two species only, and

these were sown in four different mixtures: 25 seeds of species A with 25 of B, 200 of A with 25 of B, 25 of A with 200 of B, and 200 each of A and B. Three densities were thus compared, with 50, 225, and 400 seeds sown per plot (the area was about 0.09 m^2) but the medium density was made up in two different ways.

Figure 3.4 shows the results for *Papaver rhoeas* (field poppy) when mixed with *P. dubium* (long-headed poppy). The figures on the vertical axis show the success rate for *P. rhoeas*, i.e. the chance

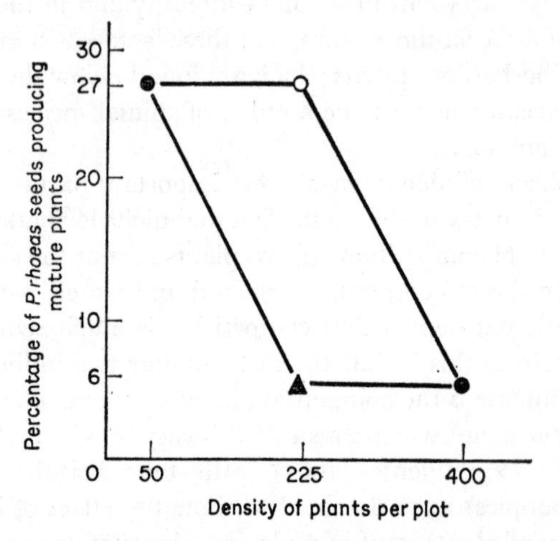

Fig. 3.4. Success rates of *Papaver rhoeas* when mixed with *P. dubium*. The horizontal axis shows the total density of the plants.
○ – 25 *P. rhoeas* : 200 *P. dubium*,
▲ – 200 *P. rhoeas* : 25 *P. dubium*,
● – equal proportions of the two species.
(Modified after Harper, 1961.)

that any seed sown would become a mature plant. Thus at the low density, with 25 of each species sown, 7 of the seeds of *P. rhoeas* survived giving a score of 27 per cent. At the opposite extreme, when 400 seeds were sown altogether only 10 plants of that species survived, which with a starting number of 200 represents a success rate of 5 per cent.

The figures for the intermediate density of 225 seeds show the most interesting feature of the results—the fact that there is a striking difference between the success rates of *P. rhoeas* in the two mixtures. With 25 of *P. rhoeas* and 200 of *P. dubium* the former still has a success rate of 27 per cent, so that by comparison with the first treatment the addition of another 175 seeds of the competitor has made no difference.

When the additional 175 are of the test species, on the other hand, it makes a great deal of difference. The mixture containing 200 *P. rhoeas* and 25 *P. dubium* showed a success rate for the former of only 6 per cent or little better than at the highest density.

The general implication of these results, which were largely mirrored not only by those from *P. dubium* in the presence of *P. rhoeas* but also by those from several other species pairs, is that the addition of many more seeds of the *same* species has a much greater effect on mortality than the addition of the same number of seeds of another species. In other words competition is much more severe between individuals of the same species than between different species. Even two species as similar as *P. rhoeas* and *P. dubium* occupy demonstrably different niches, although this type of experiment does nothing to show exactly how this is achieved. The argument is strengthened by the one atypical result in Harper's experiment. When he grew *P. dubium* and *P. lecoquii* (Babington's poppy) together he found that the two types of mixture at the intermediate density did not give the contrasting results shown by all the other pairs. It made little difference which species was present at eight times the density of the other—both had their success rates reduced more or less equally by the presence of many other plants of either species. They were occupying the same ecological niche and competing strongly with each other. The convincing final touch to this story comes when we turn to the flora for more information about the two species concerned, for of all the British poppies these two are the most difficult to tell apart. Clearly the great morphological similarity which makes identification difficult is paralleled by a high degree of physiological similarity. This is valuable confirmation of the predictive power of a taxonomy which, although based on morphology, is expected to serve the purposes of a variety of biological disciplines. That is, after all, what is meant by a 'general-purpose' taxonomy.

The importance of competition to a discussion of adaptation lies in the fact already stressed that it is universal in wild communities. When we attempt to define the environment in which a plant is living we must remember that its neighbours are a vital part of that environment. This can make it very difficult to extrapolate from cultivation experiments to wild conditions. In particular we are apt to get from our experiments an exaggerated picture of the range of tolerance of a plant. For instance in the example of lead tolerance discussed in Chapter 4, it is pointed out that resistant plants can be perfectly well grown in the garden in ordinary soil which is completely free of lead, as well as in culture solutions containing a high concentration of that metal. This suggests that they are widely tolerant of variations in lead concentration, and ought to be able to grow both on the poisonous tips from which they were collected and on the normal soil surrounding those tips.

This does not in fact happen. The tips have been there for hundreds of years and yet the highly lead-resistant plants have not spread far away from the tips, presumably because there they meet and cannot survive competition from normal plants. The same principle applies to any habitat factor we investigate in this way, as in the hypothetical experiment outlined in Chapter 1, in which a species of *Sphagnum* with an experimental pH optimum of 4.5 was found more commonly in the wild at 4.0.

The one generalization about this which does seem fairly safe is that the range of tolerance in the wild is likely to be narrower than that found in experiments. Towards the limits of a plant's physiological capabilities it is liable to grow slowly and remain small, and so any degree of competition is likely to exclude it from habitats where conditions are near these limits.

Few of the examples discussed in Chapter 4 include an experimental investigation of competition. If one plant grows more quickly or sets seed more abundantly than another it is normally assumed to be the better competitor. Clatworthy and Harper (1961) showed that this could sometimes be unwise. They found that in pure culture *Lemna polyrrhiza* (duckweed) grew more quickly than *L. gibba*, but that when they were mixed together *L. polyrrhiza* succumbed to the competition. *L. gibba* apparently had superior floating abilities which enabled it to intercept the most light and shade out the other species.

It remains to say a little more about the mechanisms involved in competition, and to widen the discussion slightly. Competition is only one form of interaction between organisms. It may be the most important in the present context, when we are considering the interactions between one green plant and another, but the same principles apply to such situations as symbiosis, predation, and parasitism. In each of these cases other organisms must be regarded as part of the environment of the one in which we are interested.

The mechanism involved in some of these situations is one of direct biological attack, as in the interaction of predator with prey. In most of them, however, and particularly in competition proper, one organism is affecting another via a change in the inorganic environment. Two plants compete for water when one dries out the soil in the rooting zone of the other. We speak of competition for light when the leaves of one plant reduce the intensity of light falling on those of another. This example differs from the first in that if one plant is taller than another the effect is not mutual but one-way, and then perhaps we should not strictly call it competition at all. It remains an important type of interaction between plants. Some of these situations are very difficult to investigate experimentally because of the clumsiness of our measuring instruments. Given big enough leaves we can measure the intensity of light falling on them reasonably accurately. It is much more difficult to show how much water is available to the root of a particular plant, and how much this may be reduced by the presence of a second plant. It is virtually impossible to assess in the field the effect of one plant on the amount of phosphate available to another, although this may be a vital factor in deciding the outcome of their interaction.

This means that we can only describe competition in rather

general terms. We can see its consequences and make important inferences about it, but we can rarely make quantitative analysis of the detailed mechanisms involved.

Plasticity and adaptation of the individual

The word 'plasticity' was used in Chapter 1, having been borrowed as a technical term. It implies that an existing form can be modified by environmental change. On the other hand plants and animals have to grow and develop from embryo to maturity, and certainly in plants those stems and leaves already in existence cannot respond morphologically to a change in the environment. It is only an indirect interaction between the environment and those genes which are operative in developing plant organs which can bring about a change in morphology. Physiological characters on the other hand are less firmly fixed during development, and there is good evidence that they can be changed by the environment even in mature tissues whose morphology is no longer plastic.

All the morphological characters of an organism are produced during the course of its development and in the last analysis this development is a chemical process. We know from work with bacteria and fungi that genes can be turned on and off, and there is some direct evidence for the control of gene action during the development of more complex organisms. There is also a great deal of indirect evidence. We do not, for example, develop eyes on the ends of our fingers. It is likely that eye-producing genes are switched on only in the appropriate environment of the developing head, while other switching mechanisms in the arms lead to the production of fingers without eyes. These various sets of genes are subject directly to the internal environment of the organism, but this internal environment may in turn be influenced from outside, and there is plenty of evidence that by this route the various characters of a plant can be altered by changes in the external environment. We might indeed expect all its characters to be equally easily altered, but this is clearly not the case. The morphology of animals is less plastic than that of plants—there is no known way, for instance, of switching on the eye-forming genes in the tissues of the fingers—but even in plants there are many characters which are extremely difficult to change. The chromosome number, for example, can only be changed in individual cells by intense irradiation or by drastic chemical treatment, and cannot be altered in the whole plant by any normal environmental influence. Many other characters are so difficult to alter that they can be used with confidence by the taxonomist; the importance of this was stressed in Chapter 1. Nevertheless it cannot be assumed that any particular feature of structure or function is immune from environmental influence until firm evidence is available.

It is a matter of common observation that some plant species are more plastic than others. Marsden-Jones and Turrill (1938) divided a number of plants into pieces and grew each piece on a different type of soil. They tested a large number of species in this way, and it is clear from their results that, for example, *Plantago major* (great plantain) was much more plastic in its overall size than was *Centaurea nemoralis* (knapweed), in this particular range of environments. The point can be made much more precisely when individual characters are considered. The classic examples here are such semi-aquatic plants as *Ranunculus peltatus* (water crowfoot), the shape of whose leaves depends on the position in the water at which they develop. Submerged leaves are finely divided into narrow segments, while those which float on the surface have broad lobes. The fully aquatic or fully terrestrial species of the same genus do not have this ability.

It must be emphasized that discussions of phenotypic plasticity should not start from some phenotype which is to be considered 'normal' and then proceed to others which are by implication less so. This is one danger of the commonly used word 'modification' in this context—it implies the existence of an unmodified form somewhere, which is highly misleading. One phenotype may be more common, or more familiar, or more successful than the others, but that does not make the others abnormal. They are all equally the product of an interaction between the genotype of the organism and its environment.

In Chapter 1, we discussed plasticity largely in the context of the difficulties it causes in the study of taxonomy or genetics, because of the resulting lack of correspondence between genotypic and phenotypic differences. We now have to inquire about its importance to the plant. It is presumably not an accident. Experience suggests that evolution has largely eliminated the accidents. The real importance of plasticity is seen when we consider the direction of response to a particular change in the environment. That response is not usually a random one, and quite different species often respond in recognizably similar ways. A reduction in light intensity causes stems to elongate and leaves to increase in size. Reduced water supply leads to the earlier production of a smaller number of flowers. In many species a reduction of daylength leads to stoppage of growth and the onset of dormancy. These changes are not an inevitable consequence of the influence of the environment on protoplasm in general, and they presumably evolved because they were of value to the species and its survival. The crucial test is whether the plastic changes observed can be shown to be adaptive to the external changes which induced them and hence enable the plant to survive better under the changed conditions.

In many cases the direct adaptive value of plastic changes still remains to be demonstrated, although persuasive suggestions can be made about particular examples. Take, for instance, the widespread evidence concerning the changes induced in plants by low light intensities. Jarvis (1964) gives some figures for oak seedlings grown under a range of intensities from full sun down to 20 per cent of that value. Table 3.7 summarizes the extreme values for three simple measurements.

No direct comparisons of fitness were made, but the changes obviously make sense in the light of what we know about plant biology. The increased leaf area would be an advantage if light was reduced to a point at which the rate of photosynthesis began to decline, and there is little doubt that 20 per cent of sunlight would be low enough for this to happen. Equally the increase of height in reduced light, which is seen in most plants, would be

Table 3.7

Response of oak seedlings to light intensity.

	20% sun	100% sun
Leaf area per seedling (cm^2)	90.0	60.0
Shoot height (cm)	15.5	11.0
Ratio root weight/stem weight	1.8	3.7

adaptive in those very common situations where light near the ground was weak but increased with height. It is part of the competitive ability of any plant to be able to grow up above its neighbours and avoid the shade cast by them, and here is a mechanism which would aid this process.

The third change is perhaps a little less obvious. There is no direct reason why a shade plant needs a smaller root/shoot ratio, in fact plants grown in shade are often less vigorous than others just because of rather inadequate root systems. The point to remember here is that the plant has only limited resources to draw on. It has a certain amount of organic matter stored in its tissues, and it manufactures a regular supply in its leaves. It is essential for photosynthesis to be kept up, and under low light conditions this may only be possible by increasing the leaf area at the expense of other parts of the plant. It can be argued that a plant with a reduced root system but a large leaf area would stand a better chance under these conditions than one with vigorous roots but too small a leaf area to manufacture a sufficient food surplus to allow it to grow.

Very many characters developed in the course of evolution look to the human observer like compromises of this kind, but this is a loaded expression derived from human affairs. The plant species evolves under the selective influence of the whole environment, so that the most viable combination of characters under conditions of low light intensity and of particular soil conditions of water and nutrient supply tends to be the most successful. The result may look to us like a compromise, but that is only because we split up the habitat into distinct factors and discuss them separately.

Some of the best examples of these apparent compromises are seen in perennial plants growing in temperate regions which have well-marked seasons, on which are superimposed short-term fluctuations of weather. Such an environment is markedly heterogeneous in time, and the compromise we read into the system is between continuous growth on the one hand and the need for protection against inclement conditions on the other. There are many environments in which no species of higher plant can grow continuously. An annual plant passes the difficult season as a seed. The perennial has to adopt a highly seasonal growth cycle which includes a period of dormancy preceded by a period of food accumulation. A great variety of detailed cycles of activity is still possible. If winter is the dormant season most of the vegetative growth is usually carried out in the spring, but there is much less agreement about the best time for reproduction. Flowering may occur in the spring, on the strength of stored food materials, or in the autumn, after the season's accumulation is over. The essential feature of any such cycle is that it must remain rigidly anchored to the correct seasons, and this is achieved by making the plant's behaviour directly dependent on the external environment. When it is considered that what we call plant 'behaviour' in this sense is largely a matter of differential growth it becomes clear that this is an important and widespread form of plasticity.

There can be little doubt that this type of plasticity is adaptive. If a small tree is prevented from going dormant by keeping it under artificial long days through the autumn, and it is then put outside, the leaves and the tips of the stems will be killed by the first hard frost. A grass can be induced to flower in the winter by controlling the temperature and daylength, but fertilization is rarely successful and any seed which forms does not ripen. There is equally little doubt that such plasticity has evolved by selection, and it will be shown under the heading of 'Adaptation to climate' in the next chapter that the products of such selection can be very precisely adjusted to local conditions.

Granted that some instances of plasticity adapt the plant to changes in the environment, how far can this generalization be extended? It is certainly too sweeping to regard plasticity and adaptation as synonymous. Non-adaptive changes can surely be envisaged, even if only in response to lethal factors. The important principle is that adaptation has to evolve, and we can predict that adaptive responses are likely to evolve as a consequence of the recurrent subjection of a population to a particular factor at a non-lethal level. There is thus every reason to expect a plant to have some adaptive response to any annual event, and to many rarer occurrences as well.

There is correspondingly little reason to expect a plant to be able to respond in an adaptive way to a stimulus it has never experienced. High-intensity irradiation has many effects on the phenotypes of living organisms but these effects cannot be expected to add up to adaptive plasticity.

There are very many examples of plasticity which would appear on casual inspection to be of an adaptive nature. It seems obvious that large leaves in the shade, or small leaves in a drought, must be advantageous to a plant. There are remarkably few careful experiments, however, which set out to demonstrate this conclusively. One practical limitation is that you cannot easily demonstrate the adaptive value of a change which is too readily reversible, if exposing the plant to a different environment immediately causes it to take on the phenotype induced by that environment. Those changes due to extensive growth are not sensitive in this way, and Whitehead (1962) was able to demonstrate that the phenotypes produced by *Helianthus* (sun flower) in a wind tunnel were effective in reducing excessive water loss and clearly had adaptive consequences for the plant.

One of the most elegant analyses of adaptive plasticity was included in the work of Björkman and Holmgren (1963) on *Solidago* (golden rod). They cultivated their material under low and high light intensities before testing it for photosynthetic efficiency. The cultivation pre-treatment was prolonged while the test was very rapid, thus avoiding the reversion effect which might have been expected with physiological characters of this kind. They found that cultivation at low intensity increased the test efficiency at low intensity. Conversely cultivation at high intensity increased the ability of a leaf to use high-intensity light in the test.

There is a connection between adaptive plasticity and the seemingly opposite principle of homeostasis, which is the ability of an organism to keep its essential physiological processes constant when external conditions change. If an organism is to survive a fluctuating environment it must clearly be able to continue those essential functions. This may range from keeping the body temperature of a mammal constant, to ensuring flowering and fruiting in a plant under adverse conditions. The link between the two ideas appears to be that an organism keeps the essential things constant by allowing the less essential ones to change in a compensatory manner. Thus sweating increases when a man goes into a hot, dry atmosphere. This is a plastic response, and it has the effect of keeping the more essential physiological activities steady through the regulation of body temperature. In the case of an annual plant very few flowers and seeds may be produced under drought conditions, but the seeds will tend to be of normal size and thus viable. By allowing great plasticity of seed number the more essential character of seed size is kept constant. In the course of evolution a plant with a small number of viable seeds would have an absolute advantage over one with a large number of ineffective ones.

4

Adaptation to habitat factors

Edaphic factors

It is convenient to set out examples of adaptive differences in terms of the type of selective factor thought to be responsible, but it must always be remembered that few of these cases have been studied in sufficient detail for us to be confident that nothing has been overlooked. One consequence of this approach is that we shall ignore examples involving quite striking differences if little is known about the selective factors involved. The justification for this is that we wish to concentrate on adaptation. It would be easy to digress into a general discussion of intra-specific variation without regard to its adaptive significance, but this has been adequately done before by the experimental taxonomists (e.g. Clausen, Keck, and Heisey, 1940).

Another introductory warning is necessary, this time a more far-reaching one with important implications for the design of experiments on adaptation. The plant does not respond to the forces which we call 'habitat factors' separately, but to a complex environment operating as a whole. Sometimes the factors we distinguish exert clearly different effects on the plant. The physiological role of phosphorus, for instance, is quite distinct from that of calcium, and when we separate them in soil analysis we are making a distinction which any plant would recognize as valid. In spite of this it is becoming clear from work with culture solutions that the whole ionic complex acts in an integrated way, so that a concentration of sodium chloride, for instance, which would be highly toxic in pure solution becomes harmless when present with a suitable mixture of other ions. In an example to be quoted later it will be seen that the toxicity of a soil cannot be predicted from its heavy-metal content alone, as there is a strong interaction with the amount of calcium present.

Other factors are even more ambiguous. We have simple instruments which measure air and leaf temperatures, relative humidity, and wind speed. The transpiration rate of a plant does not relate closely to any of these, but to a complex function of them which can be called the 'drying power' of the air. In cases such as this the 'factors' we choose to measure seem to be merely a matter of convenient instrumentation, and we cannot expect the responses of a plant to reflect the same categories.

Soil factors differ markedly from climatic factors in the pattern of their variation. Climatic factors change rapidly with time, but they may be relatively constant over considerable distances. Most soil factors are reasonably stable with respect to the life-span of the plant, but they are liable to drastic changes from point to point on the earth's surface, especially in areas of intricate geology or topography. We are thus able to study marked soil differences within a small area, whereas for some climatic studies it may be necessary to collect material from far afield.

In spite of this apparent convenience there is remarkably little satisfactory information about adaptation to edaphic factors in plants. This may be partly due to the fact that the differences to be expected are entirely physiological, while some climatic factors at least are capable of acting selectively on morphological plant characters.

Any work with soil factors usually involves the use of culture solutions and the subsequent extrapolation of the conclusions to soil conditions. This inevitably needs great care, but it is essential if we are to make any progress at all in distinguishing individual factors. A transplant experiment, in which plants are exchanged between two wild communities, merely shows their degree of adaptation to each other's habitat, without usually any hint as to which factors are crucial. It is possible to implicate soil rather than climatic factors by carrying the two soils into the glass-house and doing the experiment with the climatic factors held constant, but this still does not distinguish one soil factor from another.

The normal scheme for a culture solution experiment is first to find a solution in which all the material will grow, and then to try the effect of varying one factor at a time, in the classic manner of the formal 'scientific method'. It cannot be too strongly emphasized that this is only a start. Most of the really interesting results are produced by varying *more* than one factor at a time. The simple method is a legacy from the physical sciences, where complex interactions are much less common. It is in the interpretation of the results of multifactorial experiments that techniques like the analysis of variance come into their own. They allow the interactions between, say, two nutrient ions to be discussed in quantitative terms. The choice of particular factors to test still remains, and in practice is usually largely determined by the personal interests of the experimenter. It is very rare for material to be gathered with a completely open mind: it may in fact be very inefficient to do so. Intelligent sampling is only possible at the start if at least an embryo hypothesis is put up for testing. This usually means that one or two likely factors are in mind already, and that the composition of the first set of culture solutions to be tried is more or less determined. The technical details of how such solutions are managed are outside the scope of this account (see Hewitt, 1966).

pH and calcium

The first examples to be discussed are centred on a problem which has concerned ecologists in Britain for a long time: that of the distinction between calcicole and calcifuge species. Much is known about the physiology of such species, and we are coming to recognize a number of distinct reasons why any one species may be confined to one end or the other of the possible range of soil acidity. The reason for there being many mechanisms is that there are many factors, all varying in parallel. An acid soil not only has a low pH, but also a low content of most nutrients, especially calcium and phosphorus, and sometimes a high concentration of soluble iron, when compared with a neutral or slightly alkaline soil. Just to take one instance, many calcifuge species, genera, and even whole families, show symptoms of iron deficiency when grown in chalk soils. The resultant reduction in chlorophyll can be remedied by the addition of iron in a suitably soluble form, without making any change in either the pH or the calcium content.

Not all species are restricted to one end or the other of this soil gradient. One which is conspicuously capable of growing either on extremely acid hillsides or on chalk downs is the grass *Festuca ovina*. On either type of soil it may make up a large proportion of the sward, and yet it will be associated with completely different lists of species in the two situations. Snaydon and Bradshaw (1961) chose it for one of the first genecological experiments on adaptation to ordinary soil factors. They chose calcium concentration as the most likely selective factor, and one which was much easier to control in culture solutions than pH. Three population samples were collected from each of two extreme soil types—chalk soils with free calcium carbonate present and a pH of more than 7.5 and acid soils with very little calcium and a pH of less than 4.5. The samples were tested at five concentrations of calcium, from 5 ppm to 150 ppm in otherwise normal culture solutions. After a period of growth in the solutions the plants were harvested and their dry weights measured. The pooled results from each of the two groups of samples are shown in Figure 4.1.

It will be seen that at 5 ppm Ca the plants from acid soils grew to twice the size of those from calcareous soils, while at 150 ppm those from calcareous soils were the larger. The ambiguous results at intermediate concentrations are not significant, the differences involved being no bigger than would be expected by chance. There seems no doubt that one group of plants is better adapted to acid soils and the other to calcareous soils, but it should be carefully noted that it is the *comparative* results which are important. Neither group of plants grows better at 5 ppm than at higher concentrations. The plants from acid soils are only superior to the others at calcium levels at which all show much reduced growth.

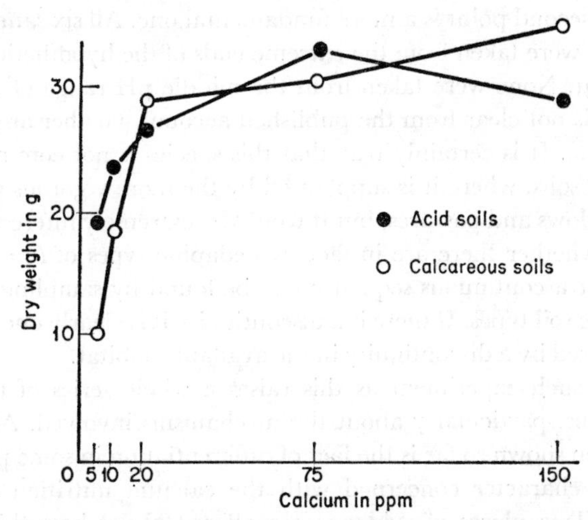

Fig. 4.1. *Festuca ovina*. The growth of contrasting types of plant at various calcium levels. The data for each type are the total for three populations from that particular soil type. (After Snaydon and Bradshaw, 1961.)

This is a clear example of the principle discussed in the section on competition in Chapter 3, that supposedly optimal conditions, as deduced from culture experiments, are usually misleading. Its superior tolerance of non-optimal conditions allows the acid-soil plant to compete successfully with the others at low Ca levels.

Any experiment can be criticized on the grounds that it should have been bigger and more ambitious. Two points need to be made about this one. It is not clear whether calcium was chosen as the factor to experiment with after prolonged search, during which other possibilities were tried but not reported, or whether it was a perspicacious first attempt. There is no denying the positive conclusions of the experiment that differentiation with respect to calcium has been found, but it would have been interesting to know whether any of the other factors in which the soils differed had also exerted any selective influence. Parallel work on *Trifolium repens* by the same authors (1962 and 1969) showed in addition differential responses to P, Mg, K, and N between groups of acidic and calcareous populations.

The second point is a more fundamental one. All six samples of *F. ovina* were taken from the extreme ends of the hypothetical soil gradient. None were taken from the middle pH range of 4.5. to 7.5. It is not clear from the published account whether any were available. It is certainly true that this species is not common on neutral soils, where it is supplanted by the more vigorous grasses of meadows and pastures, but it would be extremely interesting to know whether there are in fact two edaphic types of *F. ovina* or whether a continuous sequence can be found by sampling intermediate soil types. If there is a discontinuity it is clearly one which is imposed by a discontinuity in the available habitat.

Any such experiment as this raises a whole series of further questions, particularly about the mechanisms involved. All that has been shown so far is the fact of differentiation in some physiological character concerned with the calcium nutrition of the plant. Two pieces of evidence are offered about how this may work. The overall growth of the calcicole plants in the low calcium treatment was very poor as seen in the figures for dry weight, but the authors remark that root growth was even more reduced than these figures suggest, and that those roots which were formed were abnormal and stunted. The second observation may be connected with this. When the calcium contents of all the plants were analysed it was found that the calcicole plants grown at low calcium levels contained less calcium than the calcifuge plants. This suggests that the calcifuge plants had a more efficient mechanism for extracting the calcium from the medium, instead of having lower actual requirements.

The existence of mechanisms other than differential efficiency of calcium uptake has not been shown in *F. ovina* but work with other species suggests that the availability of iron is sometimes involved. Hutchinson (1967) showed that plants of *Teucrium scorodonium* (wood sage) from acid soils developed iron deficiency chlorosis when transplanted to calcareous soils, and that this could be remedied by the application of a soluble organic complex (a chelate) containing iron. Transplantation in the other direction, from calcareous to acid soils, gave less clear results, but there was a suggestion that the calcareous population developed abnormal roots under these conditions and that in this case it could have been due to aluminium toxicity. If this is established it is yet another example of the parallel operation of two or more habitat factors. It is quite possible for iron deficiency to be the selective factor at one end of the pH gradient and aluminium toxicity at the other. It would be equally possible for either of these to be coupled with the calcium uptake mechanism discussed for *F. ovina*, but there does not seem to be a species on which a comprehensive study has yet been made.

Heavy metals

Sometimes a single soil factor does stand out as the primary selective agent over-riding all the others. This is the case when toxic concentrations of heavy metals are present. There are waste tips from old lead, copper, and zinc mines scattered over the hillsides of Wales and the Pennines, and these contain enough residual metal to be highly toxic to plants. When the concentration is not so high that plant growth is quite impossible they become colonized with scattered plants of the grasses *Festuca ovina*, *Agrostis tenuis*, and *Deschampsia flexuosa*, and the vernal sandwort *Minuartia verna* and one or two other species. The first two of these have been studied in culture. In this case it was possible to analyse the soil of a tip and find out which metal predominated, and then test the plants for their reaction to that metal under experimental conditions. In this way the selective factor and the adaptive plant character have been closely identified.

All the published work so far has been based on the use of root growth as a measure of tolerance (Wilkins, 1957 and 1960). The rate of elongation of a root was measured in a normal culture solution, and this was compared with the rate of elongation after the addition of a standard concentration of the heavy metal. There were various technical snags such as the impossibility of adding lead to a solution containing phosphate without precipitation but these do not affect the results and the details can be found elsewhere (Wilkins, 1960). The effect of the metal on root growth was immediate. Lead, for instance, reduced cell extension in *F. ovina* to a measurable extent within less than an hour. This was clearly a very sensitive indication of the toxicity of any solution. It was found to be an equally satisfactory measure of tolerance, in that plants from lead-free soils had their rates of root growth reduced very much more than plants from highly toxic soils. Using the ratio of growth before and after adding lead as an index of tolerance, a general relationship was found between this index and the concentration of lead in the soil from which the plants had come.

One of the problems here, common to most work with soil chemistry, was that of estimating the 'available' lead in the soil. Much of the total was present as the insoluble sulphide, which would not have affected the plants. It was decided to use one of the standard extracting agents, acetic acid, to obtain a measure of the more soluble fraction such as the sulphates and carbonates, but it cannot be denied that this was a highly arbitrary procedure with a metal on which little preliminary work had been done, and that it must have been responsible for some of the discrepancies in the results. Fortunately the range of lead contents found was extremely wide, so that the question of resolving small differences did not really arise.

It was possible to take the analysis of this situation a good deal further than the calcicole-calcifuge one reported earlier. Instead of only two soil types a complete range was studied, and a reasonable correlation was found between the amount of lead in the soil and the tolerance index of the plants growing on it (Figure 4.2). There was no question of there being two distinct races of any of the grasses tested. In each case there was continuity from tolerant to non-tolerant. Merely as a point of terminology we clearly cannot refer to tolerant and non-tolerant 'ecotypes'. We have instead an ecocline, but one which may not exist as a straight line on the ground but only as something of an abstraction, in the form of a correlation between plant character and habitat factor when these are brought together from a variety of geographical locations.

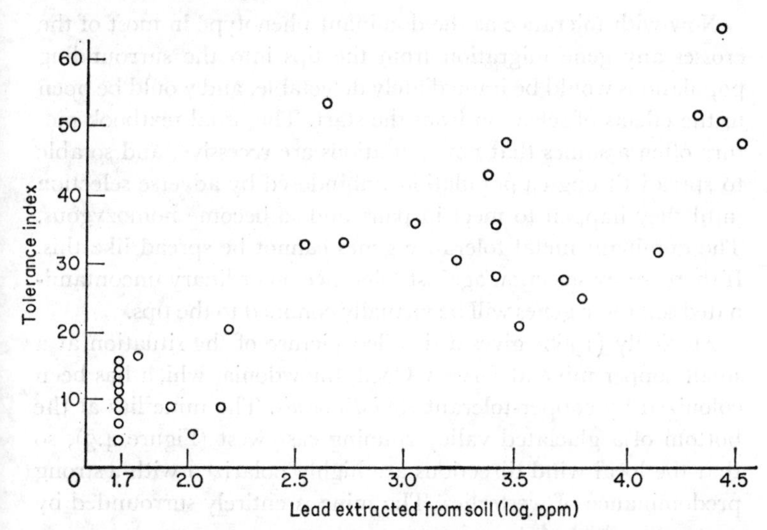

Fig. 4.2. Lead tolerance in *Festuca ovina*. Pennine population means. The tolerance of the plants is expressed as the growth rate of roots in a solution containing lead as a percentage of the rate in its absence. The soil extraction was performed with boiling acetic acid. Each point is the mean of a sample of several plants. (Data of Wilkins, D.A.)

It was also possible to investigate the inheritance of tolerance. Several generations of crosses were made in both *F. ovina* and *A. tenuis*, and in both cases it was clear that several genes were involved, as the segregations were complex and difficult to analyse. One interesting point which was also common to both species was the dominance of tolerance over non-tolerance in most crosses. Even when two extreme plants were crossed it was usual for the mean tolerance of the F_1 to be close to that of the highly tolerant parent. This made one of the observations about the distribution of tolerant plants in the wild particularly interesting. It had been noted that they were almost always confined to the tips themselves, and were very rarely picked up in the turf away from poisonous soil. This suggested that there was some reverse selection against tolerance on normal soils, although the mechanism of this is completely obscure.

Now with tolerance as the dominant phenotype in most of the crosses any gene migration from the tips into the surrounding populations would be immediately detectable, and would be open to the effects of selection from the start. The usual textbook picture often assumes that new mutations are recessive, and so able to spread through a population unhindered by adverse selection until they happen to meet in pairs and so become homozygous. The dominant metal tolerance genes cannot be spread like this. If there is any selection against tolerance on ordinary uncontaminated soil these genes will be virtually confined to the tips.

McNeilly (1968) gives a detailed picture of the situation at a small copper mine at Drws y Coed, Snowdonia, which has been colonized by copper-tolerant *Agrostis tenuis*. The mine lies at the bottom of a glaciated valley running east–west (Figure 4.3), so that the local wind directions are highly polarized with a strong predominance of westerlies. The mine is entirely surrounded by pasture in which *A. tenuis* is very common.

McNeilly took samples of both adult plants and seeds along two

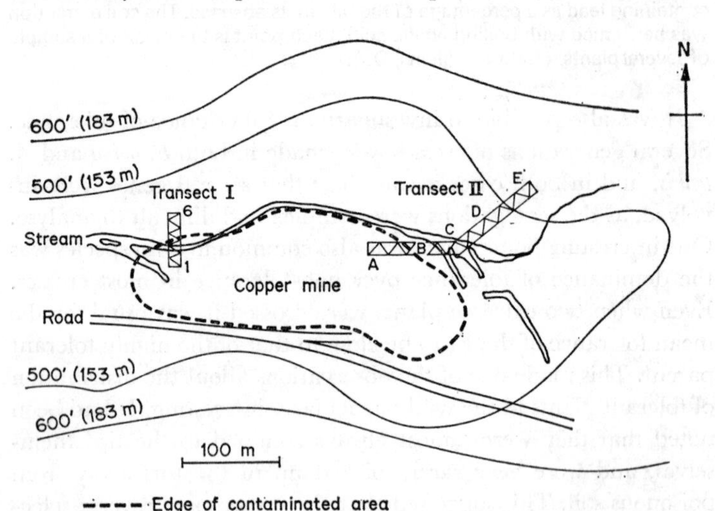

Fig. 4.3. The copper mine at Drws y Coed. 1–6 and A–E indicate the lines of transects I and II respectively.

transects (1–6, lying across the prevailing wind, and A to E, running downwind). Table 4.1 shows the mean tolerance for each

Table 4.1

Transect I	Cu in ppm	Mean tolerance of seeds	Mean tolerance of adults	Distance from previous site in metres
1	2700	39	56	0
2	2600	38	54	4
3	900	34	51	3
4	680	28	44	4
5	680	19	27	1
6	156	16	16	18
Transect II				
A	720	39	44	0
B	300	37	42	60
C	136	40	33	20
D	180	42	30	10
E	52	38	20	65

group of plants, obtained by expressing the rate of root growth with copper present in the culture solution as a percentage of the rate without it. Consider first the adult plants. It is clear that as one moves away from the mine both the copper content of the soil and the tolerance of the plants decreases. Plants with tolerance indices of below about 30 can be regarded as effectively non-tolerant. When the seed samples are considered it is clear that although the mean tolerance falls off in the same way it does so at different rates. In the cross-wind transect (I) the seed samples from the mine are less tolerant than their parents, and as one moves along the transect the tolerance of adults and seedlings decline in parallel. In the downwind transect (II) there is no decline in tolerance among the seed samples over the whole 150 metres distance. Clearly genes for tolerance are able to spread away from the tip when aided by a strong wind, thanks to wind pollination and also possibly to some slight wind dispersal of seeds. It is equally clear that at right angles to the wind, when these agencies are not effective, gene flow away from the tip is negligible. The difference between the seed and adult samples downwind of the mine is suggestive of selection against tolerance on normal soils. Further evidence pointing in the same direction is provided by the histograms in Figure 4.4. From these it will be seen that the variance of seedling tolerance is always higher than that of adult tolerance, implying that there is continuous crossbreeding between plants of different degrees of tolerance, but that selection restricts the range of variation of tolerance among plants able to grow to maturity. The selection of highly-tolerant plants on the tips is obvious enough. The selection against them on normal soil needs more study.

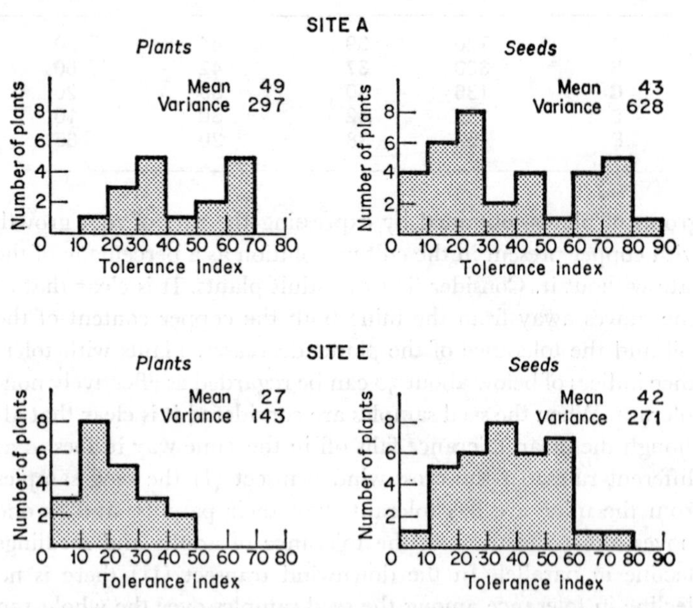

Fig. 4.4. *Agrostis tenuis*. Index of tolerance of mature plants and their seed progeny. Samples were taken from groups of plants growing at Drws y Coed.

It is here that experimental results are of help. In simple experiments with spaced plants (a spaced plant design is similar to that normally used in the vegetable garden for growing cabbages and onions), normal plants and copper-tolerant plants do not differ significantly in leaf area, leaf number, tiller number, dry weight, and net assimilation rate. Nor does there seem to be any difference in the germination rate on non-toxic soils. But it is obvious that

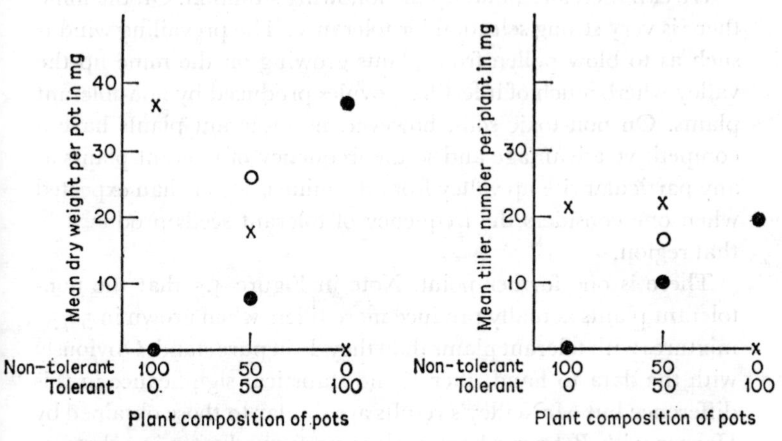

Fig. 4.5. *Agrostis tenuis*. The performance of tolerant and non-tolerant seedlings on non-toxic soils. Key: x — non-tolerant; ● — tolerant; o — tolerant and non-tolerant plants taken together.

spaced plant conditions do not correspond to natural conditions. To simulate natural conditions more closely, McNeilly (1968) transplanted groups of 42 seven-day-old seedlings into flower-pots. He ran three experiments each replicated three times. In the first experiment all 42 seedlings were non-tolerant, in the second all were tolerant, while in the third, he had a 50/50 mixture. The plants were harvested after twelve weeks. Figure 4.5 shows the results of these experiments.

Firstly there is no difference between the mean dry weight and the mean tiller number per plant in the pure stands. There is, however, a marked difference between the performance of the two types of plant in 50/50 mixtures: the tolerant plants neither producing as many tillers nor increasing in dry weight as much as the non-tolerant plants. If we can regard tiller number and dry-weight production as fitness characters, then there is little doubt that non-tolerant plants interfere with the growth of tolerant plants on normal soil and consequently the latter are less fit. Calculations show that in this experiment the coefficient of selection against tolerance on normal soils was 0.53.

We can therefore build up the following situation. On the mine there is very strong selection for tolerance. The prevailing wind is such as to blow pollen from plants growing on the mine up the valley where much of it fertilizes ovules produced by non-tolerant plants. On non-toxic soils, however, non-tolerant plants have a competitive advantage and so the frequency of tolerant plants at any particular site up valley from the mine is lower than expected when one considers the frequency of tolerant seeds produced in that region.

There is one further point. Note in Figure 4.5 that the non-tolerant plants actually produce more tillers when grown in 50/50 mixtures with tolerant plants than they do in pure stand. Obviously with the data to hand, there is no statistical significance in this difference but McNeilley's results are similar to those obtained by Harper with *Papaver* where a plant performs better in a heterogeneous biotic environment than in a uniform one, see page 61.

The work with heavy metals affords a good example of the small-scale mosaics which can be found among soil factors. Jain and Bradshaw (1966) published a list of examples of populations which had become differentiated even though very close together and thus subject to free gene-flow across the boundary. They pointed out that at Drws y Coed (Transect I) the tolerance of plants collected along a line passing from the tip to the surrounding turf dropped suddenly from high to low over a distance of only about 60 cm. In these circumstances careful sampling is essential. It is particularly important to ensure that the soil sample which is to be analysed is taken from around the roots of the plants and not from some adjacent points which looks similar but at which soil conditions may be completely different.

There is one other aspect of the lead-tolerance work which merits comment here. It was mentioned earlier that normal culture solutions could not be used because of the insolubility of lead phosphate and sulphate. On the other hand it was found that merely to use a pure solution of lead nitrate was unsatisfactory, as it was too toxic for convenience. Less than one ppm of lead was enough to inhibit growth completely in these circumstances. In the event an important interaction was discovered between lead and calcium. In the presence of a high calcium concentration it was possible to have up to 25 ppm of lead present before the root growth of a tolerant plant was much affected. This was a much more convenient figure for experimental work but the interaction had some awkward implications for any arguments about adaptation to soil conditions. It seemed clear that the toxicity of any particular soil would depend on the concentrations of both lead and calcium and the high calcium content of some of the Pennine soils in contrast with the Welsh ones made this a far from hypothetical matter. The detailed implications of this have yet to be worked out but once again we have what seemed at first to be a simple habitat factor showing unexpected interactions and again we must recognize that the plant responds to the environment as a whole.

Radio-activity

Perhaps the most unexpected example of adaptation to toxic soil conditions yet reported is that described by Mewissen *et al.* (1960) in the grass *Andropogon*. They collected seed from soils containing various amounts of uranium around the mines in Katanga and tested their sensitivity to X radiation. Seed from populations from uraniferous soils, which would have been exposed to various forms of irradiation in the wild, showed better germination and growth in the tests than did seed from normal populations. If this can be substantiated by further work it affords a most interesting example of adaptation to a habitat factor which is normally thought of as entirely deleterious to life.

Other workers have shown that the mildly toxic soil derived from some types of serpentine rock is also capable of selecting adapted genotypes in a wide range of species. There seems no doubt that examples such as these will multiply. We have to picture the average wide-ranging species as a mosaic of forms adapted to every kind of factor in which soils vary, whether this is the fluctuating concentration of a common nutrient or the unexpected presence of toxic quantities of some metal which man has mined and failed to extract completely from the discarded spoil. There are also one or two examples of adaptation to markedly different soil water regimes in the same species. The only qualification which can be suggested to the pattern of universal adaptive differentiation, and this may be merely an expression of ignorance, is that almost all the good examples concern grasses and other herbaceous plants. To some extent this reflects the convenience of working with them. There does seem to be a lack of positive results with trees, on the other hand, even when they have been looked for—although as we shall see trees readily show evidence of the selective effects of climate.

Climatic factors

The ecological effects of climatic changes are seen in the dramatic range of vegetation types encountered in a journey from the equator towards the poles, or from the coastal margins to the interior of a continental land mass. Not only does the species composition of the vegetation change but so also does its whole appearance because of the changing life form of those species. Most of the species of a rain forest are trees or epiphytes, while those of a steppe are grasses, and the differences are adaptive to a large extent to the differences of habitat.

The selective effect of climate on variation within species is equally strong. Most species are confined to a small proportion of the total world climatic range, it is true, but many are exposed to a range of conditions quite large enough to suggest that they might be physiologically differentiated.

As before it is arbitrary but convenient to group the examples according to the habitat factor involved. One of the most striking features of temperate climates is the regular alternation of contrasting seasons. When one of the seasons is unsuitable for growth there is great selective pressure to ensure that plants show an optimal pattern of seasonal behaviour, and germinate, flower and become dormant at the most appropriate times.

There are many examples of populations showing differential seasonal behaviour. Clausen, Keck, and Heisey collected *Potentilla glandulosa* from a range of altitudes in California and grew the same clones in all three of their experimental gardens. In some cases the plants were collected from near the gardens themselves, which made the experiment effectively one of 'cross transplanting' between different altitudes. An example of their results is shown in Table 4.2 in which the mean date of first flowering over a three-year period is shown for two populations and two gardens.

Table 4.2

Modification of flowering time in transplant experiments with clones of *Potentilla glandulosa*. (Condensed from Clausen, Keck, and Heisey (1940), Tables 6 and 7.)

	Flowering time at testing location		Number of plants scored	
Source of plants	Mather (sub-alpine)	Timberline (alpine)	Mather	Timberline
Sub-alpine	May 24th	August 17th	26	20
Alpine	May 17th	August 3rd	30	28

Two things are clear from this table: the alpine population regularly flowers earlier than the sub-alpine, and plants grown at Timberline flower much later than those grown at Mather. The earlier flowering of the alpine population presumably reflects the shorter growing season and the need to set seed and ripen it before the onset of winter. The difference in flowering dates of the same population grown at the two stations is a good example of behavioural plasticity, which was discussed in more detail in Chapter 3.

Daylength

Flowering is a complex process, and Clausen's type of experiment only shows the net result of a whole sequence of developmental stages, each of which may well be under the control of separate genetical and environmental influences. Of these the environmental effects are the better understood. In many species three stages can be distinguished: induction, initiation, and development. Induction is needed first and is often dependent on a period of low temperatures, referred to as vernalization. Initiation of the flowers happens at some later stage and is frequently only possible under certain daylength conditions. Finally the development which leads ultimately to the opening of mature flowers will proceed at a rate which is controlled by temperature.

More exact genecological studies can clearly be carried out on any one of these stages, but it is the second, with its photoperiodic control, which has attracted the most attention.

Before an example is discussed it is appropriate to consider in more general terms the place of photoperiodic responses in controlling the seasonal behaviour of a plant. The fundamental requirement is for some kind of calendar, which will tell the plant what time of year it is and when it needs to make preparations for the forthcoming changes in the weather. It cannot wait until the ground freezes to start becoming dormant, for by that time it is too late to accumulate food reserves and organize dormant buds. Similarly it requires months of preparation if flowering is to take place in the summer in time for the effective ripening of seed by the autumn. Very many plants have solved this problem by becoming sensitive to the relative lengths of light and dark, or less commonly of low and high temperatures. Some species can detect a change of as little as half an hour in the length of a twelve-hour day. As long as the plant can also tell spring from autumn it then has a complete calendar which is quite independent of the local vagaries of the weather. It is not misled by a late spring frost into behaving as if winter were about to set in, nor by an early warm spell into dangerously premature growth or flowering.

That is not to say that an accurate calendar is all the plant needs. The appropriate behaviour at any particular date still depends entirely on local conditions. A population descended from many generations of plants which have all been subjected to the same pattern of seasonal change will have evolved responses appropriate to those local conditions. Any individuals with seriously aberrant patterns of behaviour will have been eliminated. It is instructive to consider the detailed mechanism of this selective elimination. Let us suppose we have a perennial plant which goes dormant in the autumn in response to decreasing photoperiod. The character selected is thus the value of the critical photoperiod at which dormancy sets in. If the plant does not respond until the days are very short, it remains active into the winter, and is likely to be damaged by early frosts. If it responds to too long a day it goes dormant in the middle of the growing season while its competitors are still active. The selective factors of the habitat are thus temperature and the competitive pressure of other plants. The adaptive character, on the other hand, is not directly related to either of these but to the photoperiodic induction of dormancy.

What we have done is stress once again the integrated effect of the environment on an organism. In nature, daylength and temperature change together and there is no reason why a plant should not evolve a physiological response to one under selection pressure from the other. There may, indeed, be advantages in its doing so. In a seasonal habitat there is a premium on doing the right thing at the right time and the most efficient mechanism for doing this is the one most likely to be selected.

A simple example of the photoperiodic control of dormancy in trees is given by Vaartaja for *Betula papyrifera* (see Heslop-Harrison, 1963, p. 233). Seeds were collected from two latitudes in North America—42°N and 69°N—and germinated and grown under controlled conditions with daylengths ranging from twelve to eighteen hours. Growth was in general better under long days than short but this small effect was completely swamped by the dramatic difference between the two samples. The southern sample grew steadily at all the daylengths tested. The northern sample grew well under 18- and 16-hour days but ceased growth completely under 14- and 12-hour treatments. The effect of this in the wild would be to allow the southern population to go on growing into the autumn but to stop growth in the northern population when the effective daylight fell below a critical number of hours per day—presumably somewhere between sixteen and fourteen. The adaptive value of such induced dormancy in latitudes where killing winter frosts are frequent can hardly be doubted. Other studies have shown that this difference between a northern and a southern population is not an all-or-nothing effect. A series of populations collected from a range of latitudes show gradually changing daylength responses which can usually be related to the length of the growing season in the original localities.

An example involving time of flowering comes from the grasses of the prairies. McMillan has shown that in many different species flowering is initiated by increasing daylength in the spring and each population has its critical daylength which has to be exceeded before flowering occurs. Again a latitude series of samples shows a range of daylength responses which have the effect of ensuring that each population in the wild, flowers at an appropriate time under the conditions of its local climate. The normal pattern is that the northern populations require longer days and so flower later than the southern because of the later spring.

Both these examples involve complications in detail because of the separation already pointed out between the factor which exerts selection, temperature, and the one to which the plant responds, daylength. The correlation between these is not perfect. Daylength at any time of year can be predicted almost exactly from the latitude but the temperature is much less closely connected. A mountain top and a valley, or a west and an east coast, may be at exactly the same latitude but at very different temperatures, so that their growing seasons will not be of the same length. Plants from two such different sites would be expected to flower at different dates, and thus at different daylengths, in spite of their having come from the same latitude. Detailed information about the local conditions is essential if the results of photoperiodic adaptation are to be understood.

Light intensity

One of the few convincing demonstrations of a physiological difference between plants from sites differing in light intensity is due to Björkman and Holmgren (1963). They worked with *Solidago virgaurea* (golden rod) from woodland and from open habitats and studied its reaction to light intensity. Their experiments were complex but so careful and detailed that it is worth describing them at some length. The complications arose from the fact that they could not use one standard set of conditions in which to grow all the test plants. Because plants are plastic with respect to light intensity, the same plant grown under two different intensities will look quite different—the shaded plant will, for example, have bigger and thinner leaves and longer internodes—and these authors showed that the physiological properties were just as subject to change as the morphological ones. To match the conditions in the original habitats in this one respect, half the sample from each habitat was grown under low-intensity and half under high-intensity light in controlled environment chambers. This conditioning period lasted for several months. At the end of that time a single leaf was taken from each plant and subjected rapidly to the following test. It was enclosed in a small chamber with an air stream over it and adjustable lighting, and an infra-red gas analyser was used to obtain a continuous record of the changing amount of carbon dioxide in the air according to the rates of respiration and photosynthesis in the leaf. The rates of these two processes were estimated from the gas analysis results and this was repeated over a very wide range of light intensity.

Two figures were extracted from all this data: the rates of

photosynthesis at low and at high light intensities. These figures were compared for the two types of origin of the material—woodland and meadow—and for the two conditioning treatments—cultivation at low and high light intensity.

When considering which are the important comparisons to make among the results it must be remembered that only two combinations of conditions were realistic ones, in the sense that they resembled what the plants would have experienced in the wild. The important feature of the woodland plants was their efficiency at low light intensity after low-intensity conditioning, while in the case of the exposed plants we need to study their performance at high intensity after high-intensity conditioning. To have a complete picture of their relative efficiencies in the wild habitats we need to compare the two samples under each of these two combinations of conditions. The other two treatments, involving combinations of low and high intensity light for the conditioning and for the test, are not closely related to real life, and those results can for the moment be ignored.

Table 4.3

Rate of photosynthesis (mg CO^{2}_{3} dm^{-2} h^{-1}) of leaves of *Solidago* from sun and shade populations.
(*a*) grown and tested under low light intensity (3×10^4 erg cm^{-2} sec^{-1}) and (*b*) grown and tested under high light intensity (15×10^4 erg cm^{-2} sec^{-1}). (Adapted from Bjorkman and Holmgren, 1963, Figure 8.)

		(*a*)	(*b*)
Sun populations:	1	7	22
	2	7	24
Shade populations:	3	9	18
	4	9	17

Table 4.3 shows a brief summary of the essential results. The figures represent rate of uptake of carbon dioxide per unit leaf area. As would be expected this rate is much lower overall at low

light intensities than at high but there is an important difference between the two samples. At low intensities the shade plants have a greater uptake than the sun plants, while at high intensities the sun plants have the greater rate of uptake. As long as the rate of photosynthesis is regarded as a measure of fitness, which in these circumstances seems reasonable, we thus have a clear demonstration of a difference in physiology between two groups of plants which is likely to confer a selective advantage on them in their respective habitats in the wild.

Temperature

Temperature is an obvious factor limiting the range of many species, as can be seen from the number of introduced ornamental plants which are not hardy out of doors in Britain. There is every reason to expect temperature to be equally effective in selecting genotypes within species according to their differing requirements or tolerances. Most of the evidence here is somewhat circumstantial, however, in spite of the ease with which temperature can be controlled in experimental chambers.

It is always assumed that any variation associated with latitude or altitude is likely to be under the ultimate control of temperature even when it is manifest in such disguised form as the response to photoperiod already discussed. As an example of selection by some factor varying with altitude we have the work by Barber (1955, 1965) on *Eucalyptus*. He found that in a number of species the percentage of glaucous individuals (i.e. those with a bluish waxy surface) increased with altitude and experiments showed that the difference between glaucous and green foliage was under simple genetical control. The fact that this was a simple polymorphism on which the environment had little direct, modificatory effect allowed wild observations to be used with reasonable safety. The fact that the same cline occurred in a number of species, which presumably did not exchange genes freely, is useful circumstantial evidence that the cline is under selective environmental control. There seems no way in which parallel variation can arise in separate gene pools like this except by selection.

A second piece of evidence about selection was obtained by comparing the frequencies of green and glaucous individuals in seed samples with the frequencies among the parent trees in the same area. It was found that when a sample was taken from towards either end of the cline the numbers of green and glaucous plants arising from the seed sample was more nearly equal than in the parental population. Towards the bottom of the cline, for instance, where glaucous trees were scarce, a higher proportion of glaucous seedlings was obtained from seed. The interpretation of this seems to be that pollen was being dispersed up and down the cline, producing a regular amount of gene flow, but that at each point along it selection was operating on the seedlings in favour of the better adapted type. Glaucous plants were being discriminated against at lower altitudes and green ones towards the top. It is exactly the same argument as that employed by McNeilly in connection with copper tolerance.

The physiologist always wishes to know the mechanism behind the type of differential behaviour we are concerned with. Almost always the answer has to be that we do not know. There is a certain amount of knowledge about the metabolic basis of some varietal differences in crop plants but extremely little about the differences between wild populations. In this situation the smallest piece of information is valuable. There is such a piece, which is so far no more than a hint of what to expect, in some work on temperature tolerance in *Typha* (cattail) by McNaughton (1965). It must be presumed that physiological differences between plants are ultimately to be explained in terms of differences in the properties of their enzymes. McNaughton estimated the activity of glycollic acid oxidase in extracted chloroplasts at different temperatures and found marked differences among his four populations. In two populations from a coastal climate this activity decreased as the temperature was raised from 17°C to 37°C. In the two from inland sites the activity increased slightly over this range of temperature. (Table 4.4.) It is far too early to make dogmatic statements about the adaptive significance of this difference but McNaughton suggests that it may be connected with the fact that the coastal populations grow under low summer temperatures (perhaps 25°C) while the inland ones are subjected to much warmer summers (say 35°C). An even more important point may be that the coastal plants grow under rather constant temperatures and their enzyme activity is markedly changed by a change in the temperature. The inland plants on the other hand are subjected to widely fluctuating temperatures both from day to night and from winter to summer, and their enzyme activity seems to be buffered against this, in the sense that it changes only slightly

Table 4.4

Glycollic acid oxidase activity in chloroplast fragments from populations of *Typha latifolia*, measured at three temperatures. Figures represent microlitres oxygen/g. chloroplast/minute. (After McNaughton, 1965.)

	17°C	27°C	37°C
Maritime populations: 1	31.9	30.0	19.9
2	18.8	16.3	9.7
Inland populations: 3	8.9	11.8	12.6
4	8.7	9.9	9.2

with a 20°C change of temperature. There is an urgent need for this lead to be followed up—preferably with a much larger number of populations.

There are many other aspects of adaptation to climate. Lewis, for instance, has shown that in *Geranium sanguineum* (bloody cranesbill) the highly dissected leaves of the steppe plants are able to lose heat by radiation more rapidly than are the broader leaves of the coastal plants (see Figure 4.6). This presumably enables them to avoid lethally high leaf temperatures under conditions of drought and brilliant sunlight. There are other climatic factors still awaiting detailed study. Resistance to frost, for

Fig. 4.6. Leaves from cultivated samples of contrasting populations of *Geranium sanguineum*. A. Steppe population. B. Coastal population. (Reprinted from Lewis, *New Phytologist*, 1969, **68**. p. 496, with the permission of Blackwell Scientific Publications.)

instance, may well be of significance in montane and arctic populations. There seems no doubt that the climate, like the soil, is capable of selecting widely different responses in the complex mosaic of populations which make up a wide-ranging species.

Biotic factors

It is likely that the edaphic and climatic components of the environment exert more or less the same influence on adjacent plants of the same species. Obviously this generalization is strictly true only of uniform habitats, such as desert or savana, which can range in size from a few square metres up to thousands of square kilometres. But the generalization is relatively untrue when we consider biotic selective agents, for here we are concerned with a markedly different situation. Unlike climate, small grazing animal or plant pathogens cannot be in all parts of the habitat at the same time. Because plant and animal selective agents are not uniformly distributed, selection is not exerted evenly on all advantageous or disadvantageous phenotypes in a population. Indeed, the effects of biotic selection are dependent upon the frequency, and the density, both of the morphs being selected and of the selective agents.

Biotic selection has been studied more vigorously by the animal population geneticists than by workers with plants. Among the best examples are bird predation of the typical and melanic forms of *Biston betularia* (peppered moth), thrush predation of the colour and banding forms of the snail *Cepaea,* and the malarial parasite and sickle-cell haemoglobin in man. All of these are described in detail in another volume of this series—*The Mechanism of Evolution* by W. H. Dowdswell.

There are numerous examples of biotic selection among the host parasite relationships familiar to plant and animal pathologists. These situations have far-reaching evolutionary implications but little or no work has been done on wild populations because the pathologist works only in the laboratory or in the experimental field. Yet it is from wild relatives of cultivated species that most of the disease resistance of commercial importance is derived. Perhaps the most widespread type of polymorphism in plants is that concerned with resistance to infection by fungal, bacterial, and viral parasites. When considering the host plant it is necessary to distinguish between disease escape, disease tolerance, and disease resistance. Disease escape merely means that the plant avoids infection. Disease tolerance, on the other hand, implies that the plant responds in a passive way such that in spite of infection it can continue to grow and reproduce essentially in a normal fashion. This may be brought about by confining the parasite to some particular region of the plant. Gall formation is a good example of this restrictive process. Some parasites produce toxic substances and tolerance may evolve in the host if an enzyme, which would otherwise be inhibited, is modified or protected in such a way that it is no longer susceptible to the poisoning agent. Disease resistance may involve prevention of entry of the pathogen or resistance after the pathogen has entered the host tissue. Disease resistance after entry seems to involve antibody-antigen reactions, similar to the defensive mechanisms found in animals (particularly mammals) or the production of anti-microbial substances. Probably the most thoroughly understood examples of host-parasite relationships are those of bacteria and bacterial viruses, so we will use the interaction between the bacterium *E. coli* and the bacteriophage T_2 as an example of what is almost certainly a situation of common occurrence in host-parasite relationships.

When a dilute suspension of bacteria is spread on solid nutrient medium in a Petri dish little colonies of bacteria develop on the surface. The bacteria in each colony are all derived from a single bacterium which grew and divided independently of the others in the inoculum. The process, as far as is known, has consequences similar to mitosis and we would expect all the bacteria in any one colony to carry exactly the same hereditary information. If we now cut out one of these colonies and transfer it to liquid medium in a boiling tube, this medium will gradually become turbid in consequence of the further increase in bacterial numbers. The next stage is to add bacteriophage. The virus infects the bacteria and takes over its synthetic organelles so that the bacterium now makes viral DNA and viral protein. Thirty minutes after the initial infection the bacterial cell wall bursts, releasing approximately 200 'phage' particles. These then invade other bacteria. There may be in the tube, and there usually is, a bacterium containing a mutation for resistance to the virus. This bacterium will continue to grow and divide and while the non-resistant form will be reduced in numbers because of the bacteriophage the resistant form will increase.

By suitable means the resistant bacterium may be isolated and cultured. We can now add bacteriophage to the resistant bacteria. Among the bacteriophage there may be some which contain a mutation such that they can overcome the resistance of the resistant bacterium. These 'phages' will be able to infect, reproduce, and then burst the bacteria releasing more bacteriophage for the next cycle. Eventually nearly all the bacteria will have been killed. This new strain of the virus can be purified and stored. At this stage we have the original bacterial and viral strains together with a derived bacterial strain resistant to the original virus and a new viral strain which will attack resistant bacteria. We could now carry out the whole cycle again. In this way we eventually achieve a whole range of bacteria which are resistant to some viruses, but not others, whereas we also have a range of viruses which will attack some bacteria but not others. Genetic analysis shows that if in a bacterium there is a gene for resistance to a virus, then there is also a gene which conveys the ability to overcome this resistance in virulent viruses.

There is no doubt that this type of situation is widespread. An example among cultivated plants occurs in flax, in which some strains of the flax are resistant to some strains (races) of flax-rust fungus, *Melampsora lini*, while they are susceptible to infection by others. Flor, in the United States, has worked for a great many years with these two organisms and as a result of his studies he put forward the gene-for-gene hypothesis to explain the interaction. He found in the flax plant that resistance is usually specified by a dominant allele; while if a race of the rust could be virulent or avirulent then the allele determining avirulence was dominant. Flor found that hybrids between virulent and avirulent forms of a strain of rust segregate for pathogenicity in accordance with the number of alleles in the host which specify resistance to the avirulent form of the rust strain. This is shown most elegantly in Tables 4.5 and 4.6. The first table contains the F_2 segregation of 9:3:3:1 from the original cross of pure breeding flax varieties Ottawa 770B and Bombay. Note that Ottawa 770B is immune to rust race 24 but susceptible to race 22, while the opposite is true of variety Bombay. In Table 4.6, the results of tests for virulence of the F_2 progeny of the cross between rust races 22 and 24, show that the F_2 segregates in the ratio 9 totally avirulent, to 3 virulent against variety Bombay, to 3 virulent against variety Ottowa 770B, to one virulent against both flax varieties. We have here two highly complex polymorphic systems which are interacting probably at the chemical level of primary gene products. Flor deduced that there are complementary genic systems in the flax and the rust and that these control the reaction to rust infection. A variety of flax which carries no dominant gene for resistance is susceptible to all strains of the parasite, while a variety which contains one dominant allele for rust resistance is resistant to all strains of the rust carrying the dominant complementary allele.

Another example of biotic selection concerns the polymorphism of cyanogenesis in *Lotus corniculatus* and *Trifolium repens*. Cyanogenesis is a property of some plants whereby they evolve HCN when they are damaged. *Prunus laurocerasus* (cherry laurel) is highly cyanogenic and the squashed leaves are commonly used by collectors to kill butterflies and moths.

The production of the HCN depends upon the complementary action of cyanogenic glucosides determined by the dominant gene *Ac* and a β-glucosidase determined by the dominant gene *Li*. *T. repens* plants of the genotype *Ac Ac* or *Ac ac* contain two glucosides, linamarin and lotaustralin, whereas *ac ac* genotypes contain neither glucoside in measurable quantities. The plants can be of four phenotypes:

glucosides + enzyme (G + E)
glucosides but no enzyme (G + no E)
no glucosides but with enzyme (no G + E)
no glucosides and no enzyme (no G + no E).

Table 4.5

Segregation of F_2 plants of Ottawa 770B and Bombay for reaction to races 22 and 24 of *Melampsora lini*. (From Flor, *Advances in Genetics*, 1956, **8**, 29–54.)

Race and pathogenic genotype	Reaction* and genotype of					
	Parent varieties		F_2 plants			
	Ottawa 770B *LLnn*	Bombay *llNN*	*LN*	*Lnn*	*llN*	*llnn*
Race 22, $a_La_LA_NA_N$	S	I	I	S	I	S
Race 24, $A_LA_La_Na_N$	I	S	I	I	S	S
Number of plants observed			110	32	43	9
Number of plants expected (9:3:3:1)			109.1	36.4	36.4	12.1

$\chi^2_3 = 2.55$
P lies between 0.30 and 0.50
*I = immune; S = susceptible.

Table 4.6

Segregation of F_2 cultures of race 22 and race 24 of *Melampsora lini* for reaction to Ottawa 770B and Bombay flax varieties. (From Flor, *Advances in Genetics*, 1956, **8**, 29–54.)

Variety and reaction genotype	Reaction* of variety to					
	Parent race		F_2 genotype			
	22 $a_La_LA_NA_N$	24 $A_LA_La_Na_N$	A_LA_N	$a_La_LA_N$	$A_La_Na_N$	$a_La_La_Na_N$
Ottawa 770B, *LLnn*	S	I	I	S	I	S
Bombay, *llNN*	I	S	I	I	S	S
Number of cultures observed			78	27	23	5
Number of cultures expected (9:3:3:1)			74.8	24.9	24.9	8.3

$\chi^2 = 1.78$
P lies between 0.50 and 0.70
*I = immune (avirulent); S = susceptible (virulent).

Only (G + E) plants evolve HCN when damaged but the presence of either the glucosides or the enzyme can be determined by adding the appropriate missing substance obtained from elsewhere.

Populations of *L. corniculatus* and *T. repens* often contain both cyanogenic and acyanogenic plants (Table 4.7). How are these polymorphic populations maintained?

Table 4.7

Phenotypic frequencies of wild plants of *Lotus corniculatus* and *Trifolium repens* grown from seeds collected in different locations.
++ = glucoside plus enzyme, +− = glucoside only,
−+ = enzyme only, −− = neither glucoside nor enzyme.
(*Lotus* data from Jones, 1970; *Trifolium* data from Daday, 1954.)

	Number of plants of each phenotype				Phenotype frequencies	
Lotus corniculatus Location:	++	+−	−+	−−	Cyanogenic glucoside	β-glucosidase
Aysgarth, Yorkshire	93	2	1	0	0.99	0.98
Llanidloes, Radnor	32	50	6	20	0.76	0.35
Rounds Green, Staffs.	112	3	1	1	0.98	0.97
Blenheim Park, Oxon.	36	19	11	16	0.67	0.57
Kynance Cove, Cornwall	28	8	14	3	0.68	0.79
Trifolium repens Location:						
Aberystwyth, Cardigan	577	33	25	5	0.95	0.94
Brighton, Sussex	98	38	0	0	1.00	0.72
Cambridge	148	39	0	0	1.00	0.79
Sunderland, Durham	118	47	5	3	0.95	0.71
St. Ives, Cornwall	127	32	11	2	0.92	0.80

Daday obtained evidence showing that there was an association between the frequency of the cyanogenic form and January mean temperature (Figure 2.2) such that a decline of 1°F in temperature corresponded with a fall of 4.23 per cent in the frequency of *Ac*. For the same fall in temperature the frequency of *Li* fell by 3.16 per cent. Daday also found that there was a decline in the frequency of cyanogenic *T. repens* with increase in altitude (a negative correlation). Clearly temperature could be regarded as the selective agent because January mean temperature also declines with increasing altitude.

There are dangers with this type of work. It is likely that if you try hard enough you can always find at least one climatic factor with which variation can be correlated. Unless there is other independent evidence a correlation like that discovered by Daday can be meaningless. Fortunately Daday followed up this work and was able to show by experimental means that temperature is indeed important. He found that sexual reproduction in the (G + E), (G + no E), and (no G + E) plants was more prolific than in (no G + no E) plants under controlled warm conditions (30°C days/25°C nights). On the other hand (no G + E) and (no G + no E) plants were at an advantage in relation to (G + E) and (G + no E) under a regime of 15°C day/10°C nights. The results at high winter mean temperatures were substantiated by examining plants of known phenotypes transplanted to a coastal environment in New South Wales, Australia. Bearing in mind that July is midwinter in Australia, the July mean temperatures at the coastal location was 13°C. Daday also transplanted a total of 350 plants of the (G + E) and (no G + no E) phenotypes into an alpine region with a mean July temperature of −2°C. Under these conditions there were no differences between the reproductive performance of the two types but the (no G + no E) phenotypes were less damaged by frost.

From some other evidence, Daday suggested that it was not the character of cyanogenesis itself which was determining the response to temperature but that the locus of the cyanogenic glucoside is carried on the same chromosome as other genes concerned with plant performance.

One of the major disadvantages of working with *T. repens* is that the plant is used extensively as a fodder and ley crop and therefore it is very difficult to assess the contamination of natural populations of the plant by pollen or seeds from cultivated varieties. Hence Jones chose to work with the similar polymorphism of cyanogenesis in *L. corniculatus*, a plant rarely used agronomically in northern Europe although it is of commercial importance in North America.

Initial sampling in England and Wales revealed that there was no correspondence between the frequency of the cyanogenic form and January mean temperature. It was found, however, that several species of mollusc would preferentially eat the acyanogenic form of *L. corniculatus* when offered the choice between acyanogenic and cyanogenic plants. Observations in the field suggested that a greater proportion of plants which showed signs of grazing by small animals were of the acyanogenic phenotype than would be expected if grazing were at random and so it was suggested that this selective eating was one of the agents maintaining the polymorphism in *L. corniculatus*. Bishop and Korn (1969) on the other hand have evidence indicating that two mollusc species do not select acyanogenic *T. repens* when given the choice between the two forms.

More recent work has indicated that the phenotype itself has little or no effect on winter and summer survival of *L. corniculatus*. This means that *T. repens* and *L. corniculatus*, although possessing the same polymorphism for cyanogenic glucosides and the enzyme, differ in the way in which the genes responsible are arranged in relation to genes concerned with the fitness of the plants.

Species as ecological indicators

It has been assumed up to this point that the habitat factor behind any particular example of adaptive variation could, with some persistence, be identified. This is not always easy and we can sometimes learn a good deal about adaptation without being certain of the details of the selection process.

The simplest type of evidence in such a situation is afforded by parallel variation in separate species. The glaucousness in *Eucalyptus*, referred to earlier in this chapter, was shown by several species over the same range of altitude. These species are genetically isolated, so that a gene for a new character like glaucousness arising in one of them would not be passed on to the others. The occurrence of the glaucous character in several species at about the same altitude is strong evidence that it is being encouraged by selection.

It may help to emphasize the point if the example of *Arum maculatum* is mentioned again. The cline of frequency of spotted individuals from north to south could in principle have arisen in one of two ways. It could be due to selection by some climatic factor which acts in an unsuspected way on the spotting gene, or it could be due to the incomplete invasion of a relatively neutral gene into the British populations from a centre of origin on the Continent. We do not know which of these explanations is correct. The parallel variation in *Eucalyptus* species seems to rule out the latter type of explanation, because neutral genes, even if they arose in two species, would not be expected to give identical patterns in both of them along the same habitat gradient.

The second type of evidence for selection in otherwise obscure cases is of much more general importance and the rest of this section will be devoted to it. That is, the use of the floristic composition of the surrounding vegetation as an indication of the type of environment experienced by the plant whose variation is being studied. In the simplest case a single associated species known to have a very restricted ecological tolerance would be valuable evidence by itself. The presence of *Deschampsia flexuosa* (wavy hair-grass), for instance, or of almost any species of *Sphagnum*, would indicate highly acid conditions and if a contrasting site were sought one containing the grass *Helictotrichon* would have a much higher pH and base status. Such criteria are, after all, used by every field biologist when choosing his sample sites. He does not apply a pH meter to all the possible soils. He makes a first choice from the geological map, perhaps, and then in the field uses his knowledge of plant communities to pick the exact sites he wants.

The actual measurement of soil pH is often only necessary to convince others that he knows what he is talking about, or to investigate minor differences which are not obvious in the field.

This use of what are called indicator species is only possible when one is working in an area whose plants have been intensively studied, so that there is a fair chance of finding suitable species which can be confidently used in this way. In such a situation a genecological investigation can start along promising lines right from the beginning. If populations of a species under study were known from, say, one community containing *D. flexuosa* and another containing *Helictotrichon*, the obvious starting-point would be their reactions to pH and calcium concentration in culture.

An actual example with which one of the authors was associated concerned variation in *Potentilla erecta* (tormentil) collected from upland areas of southern Scotland. A very crude classification of the collection sites was made according primarily to the dominant species in the vegetation, as follows:

Type A. *Agrostis* spp. and *Festuca* spp. (Species-rich)
Type B. *Agrostis* spp. and *Festuca* spp. (Species-poor)
Type C. *Nardus stricta* (mat grass)
Type D. *Molinia caerulea* (purple moor grass)
Type E. *Calluna vulgaris* (ling).

The *Potentilla* was grown in an experimental garden for two years and various measurements made on it. Significant differences were found in many characters between populations from the different types of site. The problem was, could the site groups be arranged in a sequence according to the more important habitat variables in the absence of actual measurements? The number of sites was large and all habitat factors were equally interesting, so that it was hardly possible to undertake extensive physical and chemical analysis without any further guidance. Two additional pieces of floristic information were available. At each site the relative abundance of the two chromosome races of *F. ovina*, had been investigated and the *Agrostis* material collected was found to belong to two distinct species whose relative proportions could again be recorded. The site order for an increasing proportion of *A. tenuis* to *A. canina* was the same as that for increasing diploid *F. ovina:* A, B, C, E, D (the approximation to alphabetical order being coincidental). This suggestion of a habitat gradient was reinforced when the *Potentilla* data were considered; the order of the site groups for many of the characters being exactly the same as that given by the floristics. The overall picture was that the plants from sites of type A were small and those from type D large. The only possible explanation for such a parallelism of genetics and ecology seems to lie in selection. One or more habitat factors were clearly varying from A to D and selecting a gradually changing species composition on the one hand and a changing genetic constitution in *Potentilla* on the other.

The exact nature of the selective factor was never firmly established, but one very plausible suggestion can be made. The height of the vegetation was measured in the field at each site and the vegetation types had average heights which fell in the same sequence from the short *Agrostis/Festuca* turf of A to the tall stands of *Molinia* and *Calluna* of D and E. It seems likely that this difference of height was due in the first place to grazing. In the wild, *Potentilla* does not form a neat circular mat as it did in the garden but grows as long straggling shoots which spread both sideways and upwards in the surrounding vegetation. It seems likely that the grazing pressure in A would select a plant with short internodes and shoots, while in D the selection would be for plants which could avoid being smothered by the *Molinia* and climb up towards the light, i.e. those with long shoots.

A more elaborate method for showing the same type of relationship between genecological results and species composition has recently been tested (Wilkins and Lewis, 1969). The process of arranging sites in sequence according to the species present in the vegetation is properly known as ordination. A number of methods are available for carrying this out, one of the more sophisticated, when a computer is available, being the statistical technique known as principal components analysis. The effect of this is to array the sites according to the degree of floristic similarity between them, so that the two which differ most from each other should appear at opposite ends of the axis.

This technique was applied to a series of sites from which *Geranium sanguineum* had been collected for the genecological study of leaf shape by standard experimental garden methods. Figure 4.7 shows the first two axes of the principle components array,

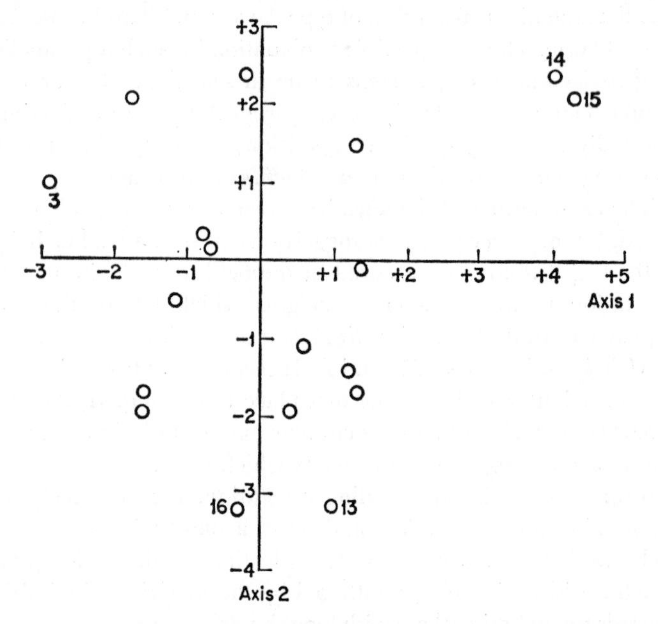

Fig. 4.7. Two-dimensional ordination of *Geranium* sites. (Reprinted from Wilkins and Lewis, *New Phytologist*, 1969, **68**, p. 866, with the permission of Blackwell Scientific Publications.)

with the reference numbers of some of the sites added. The interpretation of this is that sites 14 and 15 were very similar to one another floristically but very different from site 3, while sites 13 and 16 were again similar, but different both from 3 and from 14 and 15. In the following discussion only the first axis is considered.

The two results available for comparison were (*a*) the degree of leaf dissection of the *Geranium* population samples and (*b*) the position of the sites along the ordination axis of maximum floristic variation, measured in arbitrary units. The first finding was that these two were correlated—in other words a change of floristic composition in a particular direction was associated with an increase in the degree of dissection shown by the leaves of *Geranium*. This in itself was an indication that dissection was under selective pressure and the next step was to try and identify the factors responsible.

The site ordination could be made to yield more information by plotting the occurrence of chosen species along it. It was found that some species tended to cluster towards one end of the axis, and some towards the other, while a third group showed no particular pattern but occurred sporadically all over the diagram. When attention was concentrated on species of known habitat preference it was found that a group usually associated with dry, open, sunny habitats were clustered at the same end of the axis as the *Geranium* samples with the most highly dissected leaves. The implication of this in physiological terms is that a highly dissected leaf, when exposed to full sun, is less likely to reach lethally high temperatures than a leaf with broad lobes.

There seems little doubt that the cline of leaf dissection was selected in response to a gradient, not of air temperature in the simple sense but of that complex of factors such as insolation, air movement, and water supply which regulate the actual temperature of leaves.

This use of species presence rather than instrumentation as a guide to environmental characterististics may still seem needlessly indirect and cumbersome. There is one further argument in its favour which is worth discussing in conclusion. Instrumental measurements are frequently made at one particular time only or at best for a short intensive period of study. A plant, on the other hand, occurs in a particular place because it has survived a long and complex process of development from a seed, and during that time it has experienced a wide range of climatic conditions. One way of putting this is to say that the plant integrates all the habitat factors together over its whole life-cycle and only if the total picture is favourable is it able to survive. It is very difficult to do this instrumentally. Modern meteorological stations do record much of their data in a continuous form but it then requires an elaborate computer analysis to extract the information required and even then it may not be clear just which critical details embedded in the mass of figures are most important from the point of view of the plant.

The conclusion from this discussion is that when an ecological situation has been intensively studied for a long period it may be possible to define its environment fairly precisely in terms of measured habitat factors. When a less familiar situation is encountered, on the other hand, it may well be more informative to start by studying the floristics. At any rate this makes it less likely that any important aspect of the environment will be overlooked altogether.

5

Barriers and breeding systems

Introduction

The classical models of population genetics, showing the effects of mutation and selection on gene frequencies, usually assume random mating. This implies that any two individuals in the population are equally likely to interbreed and that gene flow in all directions is thus unhindered. This seems rather unrealistic when applied to natural populations, where many barriers to gene flow can be seen, from mere distance to actual intersterility between individuals. It is conventional to distinguish between extrinsic barriers, which are features of the environment like mountain ranges, and intrinsic barriers, which are properties of the plants themselves and include all the various forms of incompatibility. This chapter is devoted to a survey of mechanisms restricting gene flow. Polyploidy, although an important intrinsic barrier, is dealt with separately, in Chapter 6.

Extrinsic barriers

There are many topics in evolution which have always been controversial and which have not yet been dealt with to everyone's satisfaction. One of these concerns the degree of isolation which may be necessary before two populations of an outbreeding species can diverge from each other in genetic constitution. Much early work on this subject was done on closely-related species which were more or less completely isolated from one another, usually by geographical barriers. Darwin (1800) studied many animals and birds on groups of tropical islands and found clear evidence that the lack of gene flow between islands had enabled each one to produce its own characteristic species from a presumed common ancestral stock. There is no argument about this situation. When there is complete isolation the two populations are free to become adapted to the local habitat without interference from genes coming in from elsewhere.

The disagreements arise when the barriers are less complete and when a moderate amount of gene flow seems to be occurring. Many animal taxonomists take the view that species, at any rate, can only be formed when isolation is virtually complete. There have been some genetical experiments with artificial populations on the other hand, which seem to show that given sufficiently strong selection this can operate effectively in the face of very high rates of gene migration.

Some of the more obvious obstacles to gene movement between populations are imposed on the organism from outside, and are described as extrinsic barriers. Gene flow takes place at two stages in the life-cycle of a higher plant. Pollen may be carried, either by wind or by insects, from one population to another if the distance between them is not too great; or seed may be dispersed over a wide area and so add its contribution to different gene pools in the next generation. In both these ways the genetic material available to a population is increased. The extrinsic barriers to this movement concern the spatial arrangement of plants together with their mechanisms for pollination and seed dispersal, and so are commonly referred to as geographical and ecological barriers.

The distances between islands act as complete geographical barriers to many plants, as do the presence of a range of mountains, or of a sufficiently wide expanse of any other type of country in which the species is unable to grow. The pollen falls short without reaching another plant it can fertilize and the seed dispersed in the same direction does not reach a suitable habitat. When we consider smaller barriers of the same type it is clear that migration across some of them will be possible at a slow rate. Pollen may be carried an unusually great distance during a high wind, or the occasional seed may get through when carried by a wandering animal. Such low rates of movement are clearly unlikely to alter the gene pool on the far side of the barrier to any great extent. They may occasionally be important if a particular gene of adaptive value gets into a population from outside and is able to spread rapidly because of the selective advantage it conveys. This kind of thing cannot be predicted and has rarely been observed.

When barriers become less effective the situation changes. Once the distance separating two populations becomes much smaller than the average distance of pollen or seed dispersal, genes can move through what is effectively a continuous population, even though it is one of varying density. At that point selection becomes crucial. In its absence the shared gene pool would prevent any differentiation between the sub-populations. The amount of differentiation depends on the strength of selection. Much discussion of this subject has been unhelpful because of the unrealistic assumptions made about selection. It was thought for a long time that selection was a slow and largely undetectable process in the wild and that quite a small amount of gene flow might be enough to prevent its having any effect at all. It is now clear that selection can be a very powerful agent and that in some cases it can bring about rapid changes in gene frequencies even in the face of a good deal of gene migration. The example of heavy-metal tolerance discussed in Chapter 4 is a case in point. Selection for tolerance on the more poisonous tips is extreme, in the sense that non-tolerant plants cannot grow there at all. A tolerant mutant does not just have a marginal advantage but a complete one, and it does not have to compete with non-tolerant plants in order to establish. No doubt most ordinary situations involve less violent selection than this but the fact remains that it is often a strong enough force to be able to overcome much gene flow in the opposite direction.

The plants making up a population in the wild do not usually intercross in a completely random manner. There is always a tendency for more frequent crossing between near neighbours. Thus even when pollen is carried great distances it is usually so diluted by pollen from nearer sources that its chances of effecting fertilization are small. This has been studied in detail in some crop plants, where for commercial seed production it is important to know how far apart different varieties must be grown to prevent them contaminating each other. A further cause of restricted gene flow in practice is the fact that most seed falls to the ground fairly close to the female parent, even when an efficient dispersal mechanism allows some of it to travel great distances. This means that the seed falling on a lead tip, for instance, will produce a high proportion of tolerant plants, simply because most of it comes from plants already growing on the tip. There are thus two mechanisms by which a group of plants surrounded by others of the same species may still be somewhat isolated and thus able to evolve a distinctive gene pattern in response to selection. They are more likely to pollinate each other than to be pollinated from outside the group and the seed which falls in the area will mostly come from parents already growing there.

The discussion so far has been concerned with distance. When this is large the populations are said to be geographically isolated. Sometimes a distinction is made between this and ecological isolation, which is the occurrence of two populations in the same general area but in two distinct habitats. This is largely a matter of degree. Two habitats may meet along a long straight boundary, or they may be distributed in a complex mosaic. The effective isolation between them depends upon the scale of this pattern compared with the distances over which pollen and seed can be dispersed in quantity. If we imagine that one population of wide ecological tolerance is at first able to colonize the whole area, what are the chances of its evolving two distinct types each adapted more closely to life in one of the two habitats? From what has been said about isolation it will be clear that the change will take place more rapidly if there is a restriction on gene flow between the two habitats. Geographical isolation is always an aid, in this sense, to rapid evolution. It is important to emphasize that selection operates on adaptive characters only. Only those genes which offer some adaptive advantage in one habitat or the other will be altered in frequency.

Once differentiation has taken place, so that we have two sub-populations differing in their habitat preferences, subsequent migration will lead to geographical separation. Their preferred habitats will not always be adjacent to each other, so that an ecological difference, once established, often tends to maintain or even increase the isolation between the two types.

During the evolution of species growing in the same place (sympatric speciation) divergence will be more rapid if the incipient species are reproductively isolated at an early stage. On the other hand, during the evolution of species in different places (allopatric speciation) it is not necessary to keep the incipient species separate by biological methods of isolation because geography is already doing this. For example, in the genus *Gilia*, Grant (1965) has demonstrated that five sympatric species growing on the foothills in California are isolated by very strong incompatibility mechanisms while four allopatric, maritime species of North and South America can be crossed with one another with the greatest of ease.

Some older work by J. W. Gregor shows a parallel situation although it differs in detail. The *Plantago maritima* complex, generally distributed throughout Europe, shows many differences from the sub-species on the American continent. Obviously there has been great divergence in ecological requirement and tolerance and yet Gregor was able to show as a result of a breeding programme that the sea plantains of Europe and the Pacific Coast have preserved their genetic intercompatibility even though they must have been separated for millions of years.

Intrinsic barriers

Two plants which do not cross even when growing side by side must be separated by some mechanism which does not depend on distance. There is great variety of such mechanisms but a preliminary comment on their evolutionary origin may be in order. There are two distinct situations which lead to the development of intrinsic barriers to crossing. The first is the selective effect which results if gene flow is deleterious to the fitness of a population. When two adjacent populations are adapted to distinct habitats any gene flow between them is likely to reduce fitness, so that there is selection pressure in favour of a reduction in gene flow, and if the situation remains stable for long enough it is likely that some suitable mechanism will emerge. Many such examples are known between closely related species; the five species of *Gilia* were mentioned on page 117.

The second way in which intersterility can arise between species is as an incidental consequence of prolonged separation, without there being any particular benefit gained in terms of a reduction of gene flow. Presumably the barriers between the higher taxonomic levels are of this kind. A tree and a grass cannot be crossed, not because of the undoubted fact that to do so would be deleterious to both species, but because they have evolved complex physiological and morphological differences which make it quite impossible for their reproductive processes to be shared. This type of intersterility will not be discussed further because it is not relevant to the intraspecific variation with which this book is chiefly concerned.

Even the first type of barrier, where closely related populations are isolated possibly for adaptive reasons, may still involve many different mechanisms. One very important such mechanism in plants is that involving polyploidy, where a change in chromosome number immediately cuts a plant off from its relatives. If a hybrid becomes polyploid it is normally isolated from both its parent populations. If such a plant once gets established it can evolve in complete independence, with no interference from gene flow from outside. Examples will be given in Chapter 6. Other types of intersterility are distinguished according to the stage at which reproduction breaks down. This may be anywhere from pollination, which is prevented if the two species have constant, but different, insect pollinators, or if they flower at different times, to hybrid sterility, where a hybrid plant is actually produced but fails to reproduce itself. Between these two extremes there lies a great range of malfunction. The pollen may germinate but the tube fail to grow down the style. If it grows down it may not penetrate the ovule and effect fertilization. If fertilization occurs the two sets of chromosomes may not be able to coexist in a single functional nucleus and even if an embryo develops the seed may still fail to germinate because of deficiencies in the endosperm. All these mechanisms and more have been discovered in pairs of closely related species and although less work has been done at the lower levels it is clear that exactly the same barriers can arise between separate races within one morphological species.

Finally it must be pointed out that some of the breeding systems mentioned in the next section may incidentally act as barriers to gene flow between populations. If a species is largely self-fertilizing, for instance, there will be little opportunity for the transfer of genes between populations. The extreme cases of apomixis and vegetative reproduction are even more clear-cut, for if sexuality has been abandoned altogether the whole idea of gene flow is irrelevant.

Breeding systems

The simplest model of a breeding population is one in which only outcrossing occurs, rather than selfing, and in which any individual is equally likely to cross with any other. In reality such populations are rare. The randomness or otherwise of crossing has been touched on in connection with extrinsic barriers, when the importance of distance in reducing the chances of any two individuals crossing was emphasized. In a population of any size it is clear that neighbouring individuals are more likely to cross than distant ones. The other over-simplification in the model lies in the assumption that there is no selfing. When the two sexes are separate, as in higher animals, this is obviously the case but in plants there are elaborate mechanisms regulating the degree of inbreeding and outbreeding which occurs and this section is concerned with their operation.

So many organisms possess mechanisms to promote outcrossing that one must conclude that outbreeding is an advantageous character. By increasing the number of heterozygous loci, outbreeding increases the effective recombination between them.

This in turn allows a more rapid response to selection. Thus by making the action of natural selection more efficient, outbreeding itself has a selective advantage (Lewis, 1942). Mather (1940) has suggested that this advantage to the species as a whole is at the expense of some individuals of lower fitness (carrying deleterious gene combinations) and he has argued that the maximum possible amount of outbreeding may not be the most advantageous. This helps to explain why many plant species have a breeding system which is a mixture of outbreeding and inbreeding.

Monoecy is the situation whereby the plant carries unisexual as opposed to hermaphrodite flowers. Often on any one plant the flowers of one sex are produced before the flowers of the other. In *Fagus sylvatica* (beech) the female flowers are produced first, while in the vegetable marrow the male flowers are produced from the older nodes and the female flowers develop later nearer the growing point. Whereas this mechanism involves differences of time in the development of individual flowers, other mechanisms exist whereby there are differences in the time of maturing of individual structures within a hermaphrodite flower. If the anthers release ripe pollen before the stigma is receptive the flower is protandrous, whereas if the stigma ripens first so that fertilization can occur before the anthers burst, the flower is protogynous. Although these devices may be efficient at restricting self-pollination within the same flower they in no way restrict pollination between flowers on the same plant and so the system is not fully efficient as an outbreeding mechanism.

Sexual dimorphism

Sexual dimorphism exists in a wide range of plant families even though only 2 per cent of the 2200 or so species recorded in the British flora are dioecious. Sexual differentiation is the most rigid mechanism preventing self-fertilization and examples in some common plants are *Urtica dioica* (stinging nettle), *Mercurialis perennis* (dog's mercury), *M. annua* (annual mercury), and several species in the genera *Silene* (campion), *Rumex* (sorrel), and *Humulus* (hop).

It has been argued (Lewis, 1942) that the presence of a single dioecious species in an otherwise hermaphrodite genus indicates that the change to dioecy must have been quite recent. But why is dioecy so common in animals yet so rare among plants? Mather (1940) has suggested that incompatibility mechanisms are superior to unisexuality when indiscriminate mating prevails. This is because incompatibility leads to less gametic loss. On the other hand where mate discrimination occurs (in motile animals) wastage by separation of the sexes is reduced.

Self-sterility

The only fully efficient mechanisms of outcrossing in hermaphrodite plants are those involving self-sterility, where even if self-pollination takes place no selfed seed is produced. Many of these mechanisms are such that not only selfs but also certain crosses are sterile and the terms self-incompatible and cross-incompatible enable this distinction to be made.

Incompatibility can be studied from a number of points of view. Both the genetic basis and the physiological mechanism have been analysed in a number of cases, and certain plants show the additional complication of a polymorphism for flower structure, as in the case of such heterostylic groups as *Primula*.

We will consider first the outbreeding mechanisms which exist in species like *Trifolium hybridum* (alsike clover), *Prunus avium* (sweet cherry), *Oenothera organensis*, and several species of grass.

Some elegant work on the incompatibility and sterility in *Prunus avium* was performed at the John Innes Horticultural Institution over the period 1920 to 1956. Firstly Crane and Brown (1937) showed that only about 0.06 per cent of nearly 50 000 self-pollinations gave rise to mature fruit. Effectively, therefore, sweet cherry possesses a mechanism preventing self-fertilization. The incompatibility is determined by genetic factors which control pollen-tube growth and the essential feature of the genetic behaviour of incompatibility is that under normal conditions pollen cannot function in the style of a plant carrying the same factors as the pollen. This is termed an oppositional system. These factors form a multiple allelic series, the so-called S alleles, and any two of them can be carried by a given plant. Table 5.1 shows the

Table 5.1

Compatible and incompatible matings with an S allele system.

		Cross $S_1S_2 \times S_1S_2$			Cross $S_1S_2 \times S_3S_4$			Cross $S_1S_2 \times S_1S_3$	
Male Gametes		S_1	S_2		S_3	S_4		S_1	S_3
Female Tissue ♀Germ Cells	S_1	Fail	Fail	S_1	S_1S_3	S_1S_4	S_1	Fail	S_1S_3
	S_2	Fail	Fail	S_2	S_2S_3	S_2S_4	S_2	Fail	S_2S_3
		Incompatible			Fully compatible			Semi-compatible	

compatible and incompatible matings. Note that the specificity of the female parent is determined by the genotype of the sporophytic tissue whereas the specificity of the pollen, in this case, is determined by the genotype of the pollen grain.

If two plants share an S allele they will be semi-incompatible on crossing but experimentally it is not possible to distinguish between semi- and full-compatibility without carrying out breeding tests on the progeny. An individual plant will be self-incompatible because each pollen grain is bound to contain one of the two alleles present in the style; in addition it will be unable to cross with other plants containing the same alleles. The situation in detail is rather more complicated. Crane and Brown found that the different varieties of sweet cherry could be arranged in groups according to certain criteria of crossing. For example varieties in group I would not cross among themselves but would cross with varieties in groups II, III, IV etc. (Table 5.2). Of 130 356 pollinations of cross-compatible varieties made, 29 439 fruits were produced (22.58 per cent) so even these favourable crosses were only a little over 20 per cent fertile.

Some later work on *Oenothera organensis* showed that the S allele

Table 5.2

Prunus avium. The compatibility groups in the sweet cherry.
+ indicates normal fruit production; − indicates no fruit production.

	Group					
	I	I	II	II	III	VI
	S_1S_2	S_1S_2	S_1S_3	S_1S_3	S_3S_5	S_1S_4
Female parent variety / Male parent variety	Bedford Prolific	Early Rivers	Belle Agathe	Frogmore Early	Bigarreau Napoleon	Governor Wood
Bedford Prolific	−	−	+	+	+	+
Early Rivers	−	−	+	+	+	+
Belle Agathe	+	+	−	−	+	+
Frogmore Early	+	+	−	−	+	+
Bigarreau Napoleon	+	+	+	+	−	+
Governor Wood	+	+	+	+	+	−

incompatibility gene is a super-gene of at least two parts (Lewis, 1954). One part determines female incompatibility while the other part carries the male specificity. This was demonstrated as a result of the irradiation of pollen grains. If pollen taken from a plant is irradiated and used to pollinate the same plant any fruits produced must contain a mutation to self-fertility, i.e. S_1S_2 pollen irradiated gives S_1' and S_2' as rare mutations:

S_1S_2 pollinated by S_1' and S_2' pollen gives
S_1S_1' S_1S_2' S_2S_1' or S_2S_2' progeny

On examination it can usually be shown that the female specificity of the allele is intact while the male part is non-functional.

If we now carry out reciprocal crosses between S_1S_2' and S_1S_2 plants it is found that no seed is set with S_1S_2' as the female parent. The S_2' pollen is, however, accepted by the style of the S_1S_2 plant. Thus the female part is fully potent: the pollen specificity of the S_2' has been lost. In extensive work no new incompatibility alleles have been synthesized and so the origin of the multiple allelic series is still a mystery.

There is one study on the *S* incompatibility alleles in wild populations of *Oenothera organensis* in the Organ mountains in the U.S.A. Emerson (1939) counted 155 plants and collected seeds or cuttings from 115 of them. He determined that there were at least 45 different *S* alleles in these 115 plants. This is an extraordinarily large number of alleles in such a small population and two explanations have been proposed. If the population has always been a small one then this number of alleles can be explained only by postulating a high mutation rate to new alleles. Yet, as was pointed out earlier, no new alleles have been discovered in experimental material even though the techniques used were fully capable of detecting them. The more likely explanation is that the population was originally very much larger and that the 155 plants which Emerson found were merely the relics of the population.

Heterostyly in *Primula* is another example which has been analysed in detail. The genetic system here does not have the large number of alleles found in *Prunus* but is complicated in other ways. Firstly the control of the pollen specificity is sporophytic. This means that the possible matings for any one pollen grain do not depend on its own genetic constitution but on that of the parent plant, implying that some product of the diploid parental tissue remains attached to the pollen grain and reacts with the tissues of the style on which it lands. Secondly there are seven loci involved but they are so close together on the chromosome that they behave as a supergene.

The genetic mechanism is best explained by considering the

seven loci in two groups: group F containing the four concerned with style characters and group M the three concerned with anther and pollen characters. A thrum plant may then be represented as $\frac{\mathrm{FM}}{\mathrm{fm}}$, the dominant F and M alleles producing the short style and the high anthers respectively. A pin plant is represented as $\frac{\mathrm{fm}}{\mathrm{fm}}$, the recessive m and f alleles producing the long style and the low anthers. It will be seen that this leads to a 1 : 1 ratio of the two types in the progeny when the two are crossed. The other characters determined by the supergene are the morphological ones of style diameter, length of stigmatic papillae, and pollen size, and more important the biochemical characters of style and pollen compatibility. It is found that pollen will not grow on the stigma of the same plant or on that of any other plant of the same form; even if it is placed there. This means that neither the M nor the m pollen from the thrum plant will grow on the Ff style, and the m pollen from a pin plant *will* grow on Ff but not on ff. For the two types to be able to cross the thrum plant, for instance, has to be able to distinguish its own m pollen from the m pollen from a pin plant, and this entails, as already mentioned, sporophytic control.

Pamela Dowrick (1956) has strong indirect evidence that crossing over between the F and M parts of the supergene in a thrum plant would lead to a mixture of characters in the progeny after crossing with a normal pin plant. Two new types would be formed, $\frac{\mathrm{Fm}}{\mathrm{fm}}$, with short styles and low anthers, called the 'short homostyle', and $\frac{\mathrm{fM}}{\mathrm{fm}}$, the 'long homostyle' with long style and high anthers. The latter plant is of interest because of its regular occurrence at high frequency in certain natural populations of *Primula vulgaris*. It has not only the style length but also the style incompatibility characteristics of a pin plant, coupled with the anther position and the pollen incompatibility of a thrum. This makes it self-fertile. In terms of the breeding system it thus only requires this crossover to convert a completely outbreeding species into one which can inbreed. There are some circumstances in which inbreeding rather than outbreeding confers a selective advantage. If an inbreeding mechanism evolves whereby a particularly advantageous genotype can be perpetuated, the change in the breeding system can itself be regarded as adaptive. A change in the breeding system from regular outcrossing to habitual inbreeding will result in a plant becoming reproductively isolated from similar plants growing in the same population. When this occurs it is likely to affect only a few individuals in a population and therefore unless the inbreeders show a marked selective advantage they will remain relatively few in number.

There are many examples of change in the breeding system but few better than in upland cotton. The outbreeding upland cotton spread from Mexico to the U.S.A. during the nineteenth century and in the intervening time two forms have appeared which differ in their breeding system. In the Eastern states, where bees are plentiful, cotton is regularly outcrossed, whereas in the Western states outbreeding is uncommon, perhaps due to a shortage of pollinating insects. In addition to this change in the breeding system the form in the Western states has a smaller size, smaller number of fruits, and a shorter flowering season than the Eastern form. When inbred, the Eastern form shows inbreeding depression while the normally inbred Western form does not. This has been interpreted by suggesting that in the West those individuals with a genic composition suitable for the change from outbreeding to inbreeding have been selected. Note however that the Western form is not a complete inbreeder and so the change from outbreeding to inbreeding was by no means absolute. Thus there was time for the readjustment of the genic composition of the population to occur before the inbreeding resulted in the loss of the population altogether (by depression) or in complete homozygosity. This last point is particularly important, for in those species which regularly inbreed it has often been found that there are special devices for maintaining heterozygosity.

An extreme example of such a device is observed in some

species of the genus *Oenothera* (evening primroses). Here there may be a balanced lethal system associated with chromosome interchange such that homozygotes die. Indeed in some species, e.g. *Oe. lamarkiana* and *Oe. biennis*, all fourteen chromosomes are involved in interchanges and only two types of gamete are produced (see for example Lewis and John (1963) for the genetical consequences of interchange heterozygosity). This means that half the selfed progeny will be completely homozygous and inviable while the other half will be heterozygous and, apart from mutation, will be identical in genotype both with the parent and with each other. In this way the parental genotype is perpetuated essentially unchanged. Surprisingly, in another species of the same genus, *Oe. organensis*, there is an elaborate incompatibility system and usually the fourteen chromosomes form seven bivalents.

Another, and much more common way of preserving an advantageous genotype is by vegetative reproduction. The formation of tillers by grasses, runners by strawberries and white clover, and rhizomes by potatoes and stinging nettles are all familiar examples and one presumes there must have been, at times, very strong selection favouring plants which had the capacity for producing vegetative propagating organs. Some clones produced in this way can be very large. Harberd (1961) has evidence that the same genotype of the grass *Festuca rubra* can be found in an area 220 m across (see Chapter 6).

How frequently a change in the breeding system has been forced upon plant species, as it was in upland cotton, cannot be determined but there is no doubt that there can be advantages in fluctuations in the breeding system. It is clear that whole groups of plants benefit from the combination of sexual and vegetative reproduction and as we have seen in such plants as *Primula* it may be possible to alternate between outbreeding and inbreeding.

Homostyle plants, which are known to be self-fertile, occur at low frequency in many wild populations of *P. vulgaris*. These do not appear to multiply at the expense of the normal plants and there is no evidence that they have any particular advantage over them. It might even be expected that they would be at a disadvantage, because the result of selfing a normally outbreeding species is frequently to produce a reduction in general vigour. It seems that in most *Primula* populations there is a selective pressure in favour of outbreeding, via the normal incompatibility system.

Two populations are known in southern England which are unusual in having a high frequency of long homostyle plants, which may in some local colonies attain as high a figure as 80 per cent. It might be supposed that a particularly successful form of homostyle had arisen and that it was on the way to ousting the normal plants, but this does not appear to be the case. These populations have now been studied for as long as thirty years and the homostyles have not increased their numbers in that time, suggesting that they do not in fact show any particular advantage or disadvantage today in competition with the normal heterostyles. We can only suppose that at some period in the past the homostyle had a temporary advantage in these two areas, which enabled it to increase its numbers to such a marked extent. Can a possible mechanism be suggested? The one important way in which a homostyle differs from a heterostyle is the ease and success with which it can be self-pollinated. Unfortunately the evidence about the amount of selfing which actually takes place in the wild is conflicting, with estimates ranging from 20 per cent to more than 90 per cent. The anthers and stigma grow in close proximity and pollen is readily transferred from one to the other, but on the other hand the stigma appears to ripen before the anthers open and this gives foreign pollen a chance to get in first. If selfing were a frequent occurrence it could be suggested that the homostyles would have a marked advantage if for any reason insects were scarce, as in those circumstances they would be able to set seed much more reliably than the cross-pollinated heterostyles. This could be one speculative explanation for the original spread of the homostyle in these two populations. More details of this interesting puzzle can be found in Ford (1971).

Another character which can have a marked effect on the breeding system is time of flowering, simply because of the obvious fact that two plants can only cross if their flowering times overlap. Before the importance of flowering time in any particular species can be investigated we have to consider a number of other aspects of the breeding system. For instance we need to know whether the plant is self-fertile or whether it is an obligate outbreeder. We have to answer questions such as the following. Does the plant produce one, several, or many flowers? If the flowers are solitary, for how long are they open? In multi-flowered plants, are they produced in succession or are they all open at once? Given adequate pollen does each flower produce one or many seeds?

These questions lead on to other points. (1) The production of a succession of flowers should enable an individual to cross with a variety of other plants of that species with which it overlaps in flowering time. (2) For any one pollination by insects, can one say anything sensible about the source of that pollen? That is, how many plants provided the pollen which was deposited by the pollinating insect? (3) Plants with the same flowering time, which produce solitary flowers, may well have similar genotypes so that crossing is almost equivalent to selfing.

It is now clear that these problems are very complicated, but they are being tackled by Lawrence (1965) and his colleagues in Birmingham. Initial results with *Papaver dubium* suggested that there was much inbreeding in wild populations of this plant. Later work, using refined techniques and material from different populations, revealed that the initial explanation was too simple. They found that the degree of inbreeding is highly variable and the consequences of this are being investigated.

Experiments with *Arabidopsis thaliana*, which Clapham *et al.* (1962) describe as automatically self-pollinated, show that outcrossing occurs more frequently than would be expected from the morphology of the flower. Using a major-gene marker (presence/absence of hairs on stems and leaves) an estimate of 1.73 per cent outcrossing has been obtained by examining over 10 000 progeny from suitable open pollinated plants. This level of outcrossing may seem trivial but it is enough to maintain a not-insignificant degree of heterozygosity in wild populations.

Hence it is unwise to guess at the breeding system of a plant merely from the external appearance of its flowers or from its ability to set seed in a glasshouse in the absence of pollinating agents. The only satisfactory way of establishing it is by breeding experiments in conditions where possible pollinators occur.

Asexuality

Up to this point in the discussion of breeding systems we have dealt only with mechanisms associated with formal sexual reproduction. In certain groups of plants other methods have been developed which are alternative or additional to sexual processes. Some of these still involve the fusion of two haploid nuclei, although they are no longer derived from male and female gametes, while in others meiosis and fusion have been abandoned altogether. Table 5.3 sets out the consequences of some of the possible sequences of events. Three of these sequences lead to the perpetuation of the parental genotype: vegetative cloning, parthenogenesis and one type of apomixis—and hence they are useful for maintaining immediate fitness, though at the expense of long-term flexibility.

Sexual reproduction on the other hand maintains flexibility although the average fitness of all the individuals is below the maximum. The majority of plant species show a balance between these extremes so that there is a certain percentage of outcrossing among plants which appear to be perpetual inbreeders or apomicts. Stebbins (1957) has pointed out that apomixis has not developed in self-fertilizing plants; the sexual ancestors of habitual apomicts are dioecious or self-incompatible, or they possess other mechanisms for ensuring cross-fertilization. In addition, apomixis is not uncommon among plants of hybrid origin and among polyploids. Why should this be so? One suggestion is that apomixis is an escape from sterility. A plant isolated geographically or genetically from its relatives is not in a position to reproduce by seed unless there is a breakdown of meiosis leading to the

Table 5.3

Sexual and non-sexual modes of reproduction.
(Note that some authorities use the term apomixis in a wider sense to include both parthenogenesis and vegetative reproduction.)

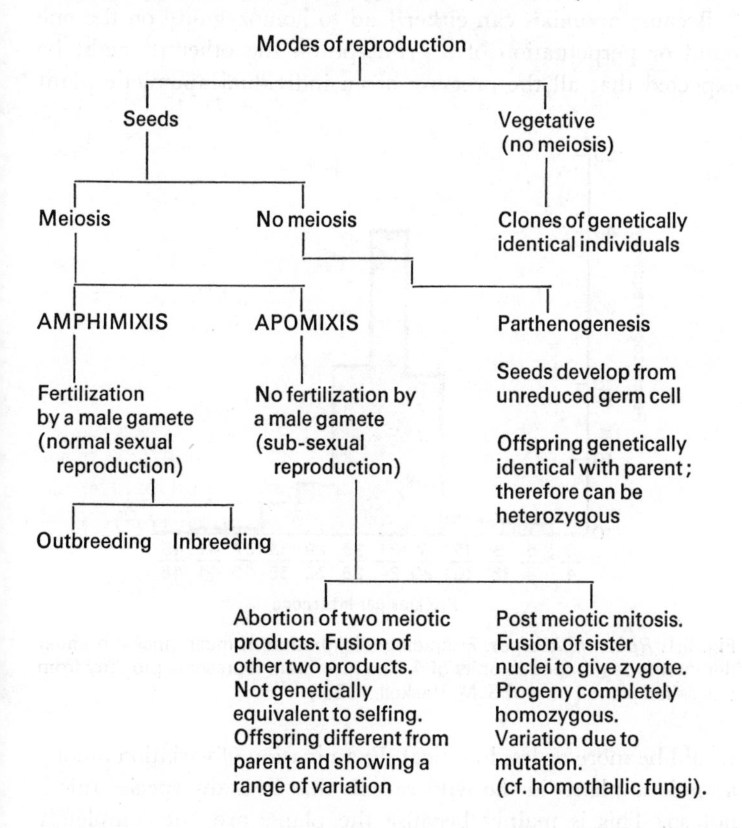

production of a diploid nucleus within an embryo sac. Table 5.3 gives some examples of how this may come about and it is not difficult to appreciate that in an effectively sterile individual any abnormal process which leads to the production of fertile seed will be at an enormous advantage. There will also be selection favouring relaxation of the genetic control of the meiotic process.

It will be pointed out in Chapter 6, in the section on polyploidy, that an alternative escape for hybrids from this impass of sexual sterility is by chromosome doubling but even this process may not be sufficient in itself if the original parents were self-incompatible.

Because apomixis can either lead to homozygosity on the one hand or perpetuation of a genotype on the other it might be expected that all the progeny of an individual apomictic plant

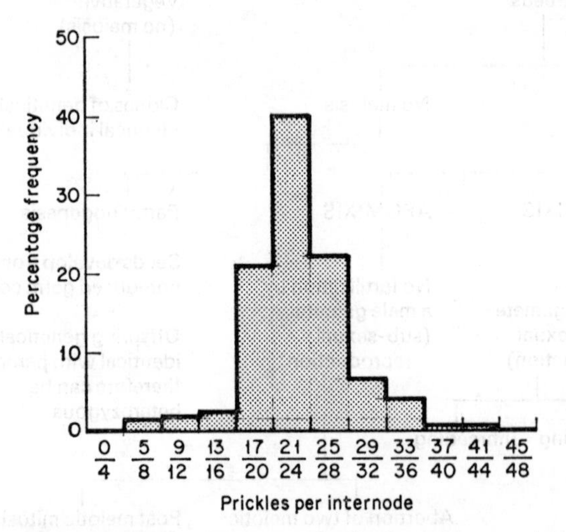

Fig. 5.1. *Rubus nitidioides.* Frequency distribution of mean prickle number per internode grouped in units of 4. The histogram represents progeny from one selfed plant. (Data of G. M. Haskell, 1953.)

would be more or less identical. Examination of variation among apomictic plants in the wild reveals that in many species this is not so. This is mainly because the plants are not completely apomictic. In the raspberries and blackberries (*Rubus* spp.) for example both parthenogenensis and breakdown of meiosis occur in several species and all this is superimposed upon a normal sexual process (see Haskell, 1961). Stimulation by pollination is required however before any apomictic seed can be set. Figure 5.1 shows the frequency distribution of mean prickle number per internode, grouped in units of four, in the progeny produced by 'selfing' one individual of *Rubus nitidioides* (Haskell). It is clear that there is great variation in the family and yet it is not possible to say precisely what proportion and which individuals were produced by amphimixis and which by apomixis.

6

Clones and polyploids

Clone formation

It is well known that many plants can maintain themselves for long periods without sexuality, by means of what is loosely known as 'vegetative reproduction'. This may entail nothing more specialized than a creeping stem able to root at the nodes, as in many grasses, or it may involve elaborate structures such as bulbs, fungal spores, or the gemmae of many bryophytes. Some of these methods lead to a dense stand of plants of a single genotype, while others lead to the wide dispersal of that genotype in the form of a large number of isolated individuals.

This has many implications for theories of population genetics. Such theories are usually based on *Drosophila*-type organisms, in which the individuals are clearly recognizable, have a limited length of life, and, except in certain exceptional cases, are all genetically different. Any discussion of the ratio between two alleles in a population, on the other hand, becomes much more complicated if individuals are able to reproduce without any sexual process intervening, so that the same genotype can occupy a large area of ground and in principle perhaps live for ever if conditions remain favourable. It is clearly possible that one or two favoured individuals might occupy all the available space in a given area, so that an apparently large population really contained only one or two distinct genotypes.

If this were to happen it would no longer be correct to think of the population as a product of the continuous selection by the environment of a steady stream of new genotypes. The genetic composition of such a population could well depend on the accident of which seeds arrived on the site first. The founder members, once established, could in a closed community prevent the incursion of any more seedlings. Even if potentially superior genotypes came along they would have no chance of replacing the founders. One frequently sees stands of such plants as *Pteridium* (bracken), *Phragmites* (reed), and *Spartina* (cord grass) which look very much like single plants which have spread out from a central focus but the only critical work on the subject seems to be that of Harberd (1961) and he found a more complex situation than that suggested above.

Harberd sampled a creeping grass, *Festuca rubra*, by collecting tillers from a wide area of mixed grassland, and he then grew these tillers as spaced plants in an experimental garden. After they had been growing for a year or more it was possible to walk round and match plants which looked alike, and two experiments were devised to try and consolidate this subjective impression that some plants had been collected more than once from separate points in the wild. It is an interesting feature of the situation that proof was inherently impossible. Harberd wished to show that pairs of plants were genetically identical, while the only conclusion available from any statistical analysis is that such plants are not significantly different. This is a very general point of scientific method, which Harberd clearly recognized: differences can be tested and confirmed, while similarities can not. In spite of this it was possible by careful experimental design to establish a high degree of probability in favour of the idea that single genotypes had spread widely and been sampled more than once.

The first experiment involved the deliberate cloning of the material. Twelve plants were selected which, although they had been collected separately in the field, looked extremely similar in cultivation, taking into account as many characters as possible in assessing similarity. Each plant was divided into ten ramets and these were planted out in a randomized layout, At the end of the growing season various measurements were made on the trial and these were then subjected to an analysis of variance. The critical comparison was between the variance among the twelve original plants and that among the deliberately manufactured ramets within these plants. In no case was the variance between the plants significantly greater than that within. This contrasted markedly with all Harberd's experience with garden trials of both *F. rubra* and *F. ovina*. Whenever distinct genotypes were included in a trial the differences between them were always highly significant for most of the characters measured. There seems little doubt that he was dealing with two quite different levels of variance, and that in the 'cloned clone' trial the separately collected plants which looked so similar to the eye were in fact of the same genetic constitution. The more characters Harberd examined and tested the stronger this impression became.

The second experiment was more restricted in its material but at the same time more searching. *F. rubra* has an *S* allele incompatibility system. If the number of *S* alleles is very large the chances of any two random plants being unable to cross are very small. Studies on other species with similar systems suggest that the number of alleles might in fact run into hundreds and in practice it is rare to find two different plants which cannot be crossed. Harberd made use of this by attempting a very large number of crosses among his plants and recording the amount of seed set in each case. He also tried to self a number of plants for comparison. He found that as expected the selfed plants produced little seed, while most of the crossed plants produced a great deal. The significant observation was that in those cases where two separate plants set very little seed on crossing they always looked morphologically identical. This strongly suggested that they were in fact parts of the same plant, although they had been collected separately in the wild. The chances of picking up the same incompatibility alleles a second time are small in any case. The chances of finding it linked with the same morphological features as well are negligible.

When Harberd's results were translated back to the map of the original collecting sites it became clear that some of his plants were very large. In one intensively sampled plot 10-m square few genotypes were found to be confined to small areas of it. Most of them gave the impression that only a portion of their distribution had been included. This was amply confirmed by a wider survey, in which one genotype was found scattered over a patch 220 m long. Not only was it widely scattered but in many parts of its area it made up from one-quarter to three-quarters of all the tillers of *F. rubra* collected.

The conclusion which seems inescapable from this is that the big plant was very old. There seems to be no mechanism of apomixis in the species by which a single genotype can spread through normal seed dispersal and it must be assumed that ordinary tillering was responsible for the spread of these plants. Even under the most favourable conditions this could not have taken less than 400 years, and there is every chance that it might have required 1000 years. Harberd points out that this raises several interesting points about adaptation. It calls in question the general view that plants are so critically adapted to their environments that a small change in conditions would cause them to be ousted by better adapted competitors. Certainly in 400 years that Scottish hillside must have experienced a wide range of management regimes, a moderate degree of climatic change, and probably a good deal of variation in the associated species. The other aspect of this same point is that the area of ground over which the large plant was growing was not particularly homogeneous, as it included a variety of grassland types of quite different composition. It was thus a field example of the 'one genotype in several environments' type of experiment referred to elsewhere. The fact that this same genotype was abundant and had survived for a long period under such a variety of soil conditions was again evidence of its wide ecological tolerance.

In spite of this impressive evidence of permanence on the part of one or two genotypes they did not, as postulated earlier, make up the whole of the population. There was a great mixture even in the intensively sampled 10-m square site. Out of the 1481 tillers taken at regular intervals across the site 51 per cent were of the 'big' genotype previously described. Eleven other genotypes contributed ten or more tillers each. At the other end of the scale no less than 150 apparently different genotypes were represented once only. The picture is thus one of a combination of a few abundant genotypes with a large number of rare ones. There is no evidence as to whether the rare ones are new arrivals or less successful veterans. The one thing which is quite clear is that we cannot with any confidence talk about an average plant in such a community and the sampling problems remain very severe.

Polyploidy

We have considered elsewhere the mutation of genes and the origin of new variation by this means and we have given some indication of the speed of the processes of change. Mutation of whole chromosomes can, however, be so fundamental that a new type of plant arises almost instantaneously. This is particularly true when the chromosome number is doubled, as can happen when chromosome duplication during mitosis is not followed by nuclear division. As we have seen earlier, there are still arguments going on between taxonomists and geneticists over what constitues a distinct species. It is obvious that if a complete biological barrier to crossing arises between two groups of individuals then no exchange of genes can occur. If this barrier is set up rapidly, as happens on chromosome doubling, we could have two groups of plants with essentially the same morphology which cannot interbreed. 'Two species' says the geneticist; 'chromosome races of the same species' says the taxonomist.

One can argue that the form of a plant matters little as long as it can reproduce for, in the last analysis, it is the leaving of progeny which determines whether or not an individual has been a biological success. From this point of view the disagreement is merely one of terminology, involving the definition of a species.

Chromosome doubling entails the addition of whole sets of chromosomes and mutations of this type give rise to polyploids. The terms haploid, diploid, tetriploid, traploid, etc., are applied to plants whose nuclei contain one, two, three, four or more sets of chromosomes. On the other hand, errors of mitosis involving single chromosomes can give rise to addition or loss of these chromosomes, a situation termed aneuploidy. The nomenclature here also gives a good description of the type of aneuploid for a diploid missing one chromosome is a monosomic diploid ($2n = 2x - 1$, where n is the gamete number and x is the basic number of chromosomes) while a diploid with an additional chromosome is called a trisomic diploid ($2n = 2x + 1$) because one of its chromosomes is represented three times.

How do polyploids arise? There are several theoretical models of the origin of polyploid plants and at least one method can be simulated by use of the drug colchicine. When growing seedlings are placed in a dilute solution of the drug it is absorbed into the tissues where it interrupts the formation of the mitotic spindle in dividing cells. Consequently chromosome duplication is not followed by normal segregation of the products of the duplication and in the absence of cell division the 'daughter' cell contains double the somatic number of chromosomes.

Two types of polyploid can be recognized, the distinction being based upon the relationship between the four sets of chromosomes, though it is often not possible to place any one plant or group of plants into one or other category. In autoploids the initial doubling occurs in a fully fertile diploid individual and, because there are four homologous sets of chromosomes (four genomes), any one chromosome has full pairing potential with three others at meiosis. As a result there are five possible pairing relationships within the group of four homologous chromosomes: two bivalents; a trivalent and one univalent; a quadrivalent; a bivalent and two univalents; or four univalents. The last four usually give rise to unequal segregation of chromosomes at anaphase I of meiosis, leading to lowered fertility. Regular pairing in the tetraploid would give rise to greater fertility and so would be favoured by selection. It is not surprising, therefore, that autotetraploids seem to have made little contribution to the evolution of flowering plants.

At the other extreme are the polyploids formed by hybrids. One consequence of interspecific crossing is that the hybrid is often sterile. If the differences between the parents are considerable then errors in meiosis may be expected. Thus zygotene pairing may be unsatisfactory, so that either bivalent formation is incomplete or the bivalents fall apart at diplotene. In addition there may well be differences in chromosome number between the parents and thus lack of pairing partners will give rise to the formation of univalents rather than bivalents, and the meiotic products will carry an excess or a deficiency of chromosomes. These and the other upsets which can occur will give rise to unfertile gametes or, if the gametes are fertile, to non-viable zygotes. In such sterile hybrids, therefore, there is strong selection for a state whereby regular pairing can occur and this can be achieved by fortuitous doubling of the chromosome number. In consequence the hybrid becomes completely homozygous with respect to each of the chromosome sets it carries but, on the other hand, it is permanently heterozygous between the two genomes. Tetraploids formed by the doubling up of chromosomes in a hybrid are allotetraploids or, when the meiotic pairing is strictly as bivalents, amphidiploids.

There are other consequences of chromosome doubling in a species hybrid which must be considered. In self-sterile species, unless doubling occurs in more than one of the progeny of a hybrid cross the individual amphidiploid will not be able to reproduce sexually. This puts a restriction on the species which can become polyploids for it is extremely unlikely that mutation both to polyploidy and self-fertility would occur simultaneously. If, however, the new sterile polypoid can reproduce vegetatively, a factor of time is introduced allowing a greater chance for a mutation to self-fertility to occur.

Between the extremes of true autopolyploidy and allopolyploidy there is a whole range of polyploid types. Segmental polyploids have chromosomes part of which have homology with three other chromosomes while the other parts have homology with only one other chromosome. It is the difficulty of analysing this situation which makes categorization of polyploids so hazardous.

Polyploidy in wild plants

Genetical evidence suggests that *Lythrum salicaria* (purple loosestrife) regularly forms quadrivalents in at least two chromosomes, because two unlinked characters show tetrasomic inheritance giving ratios of 35 : 1 in the F_2.

There is cytological evidence that *Lotus corniculatus*, on the other hand, very rarely forms quadrivalents or trivalents but this species shows tetrasomic inheritance for seven characters which are not genetically linked. The size of the chromosomes and the frequency of chiasma formation are important here. Small chromosomes cannot, for mechanical reasons, form a large number of chiasmata but a chiasma frequency in excess of one per bivalent is required for synapsis to be maintained in quadrivalent. If, therefore, the chromosomes of *Lotus corniculatus* are smaller than the critical size needed for more than one chiasma per bivalent normally to be formed then it follows that regular bivalent or trivalent formation will result. This appears to be the case, so cytologically the plant behaves as a diploid while the mode of inheritance of all but one of the characters so far studied is tetrasomic.

These two species are representative of a small number of known autotetraploids of natural origin. By far the majority of polyploids are classed as allopolyploids, the judgement being based almost entirely on cytological evidence. The formal genetic analysis of tetraploids is by no means easy and it was not until Sir Ronald Fisher became interested in the problem of tristylic incompatibility in *Lythrum salicaria* that any logical breeding programme evolved. Because it is so much easier to look at chromosomes in meiotic prophase than it is to carry out the necessary crosses for at least two generations, most polyploid plants are classified as allopolypoids unless there is a high frequency of quadrivalent formation. But it has already been pointed out that quadrivalent formation is related to chromosome size and chiasma frequency and so the whole argument is far from satisfactory.

It is as difficult to establish the origin of an allopolyploid as it is of an autopolyploid. The process is essentially a piece of scientific detective work, the sequence of events being built up in retrospect.

Of course there is the possibility of repeating the process when the putative ancestors are still in existence. Without any doubt the most satisfactory account of the evolution of an allotetraploid relates to the origin of *Spartina* x *townsendii* agg. even though the original account, which has been widely quoted, was incorrect in certain essential detail. Recent work by Marchant has cleared up a number of points and the following is a résumé of the present situation as described by him.

Spartina x *townsendii* agg. is a general name which includes both the sterile F_1 hybrid derived from introduced *Spartina alterniflora* and native *S. maritima* and the fertile amphidiploid which arose from it. Examination of the meiotic and mitotic chromosomes of several species of *Spartina* give good evidence that the basic number is 10 so that *S. maritima* with 60 and *S. alterniflora* with 62 chromosomes are both hexaploid with one chromosome being represented eight times in *S. alterniflora*, i.e. $2n = 6x + 2$, an octosomic hexaploid!

Historical evidence places the appearance of the *S.* x *townsendii* hybrids in Southampton Water at about 1870 and this was some forty years after the first record of *S. alterniflora* in the River Itchen which flows into Southampton Water. It is likely that *S. alterniflora* was accidentally introduced among shipping ballast.

Morphological and ecological investigations strongly support the theory of the hybrid origin of *S.* x *townsendii* and the cytological evidence is nearly conclusive. The sterile F_1 hybrid has $2n = 6x + 2 = 62$, while the amphidiploid has $2n = 12x + 2 = 124$, or 122 or 120 in three chromsome races. This in itself is not such a tidy story as that originally put forward by Huskins, for the expected numbers would be 61 and 122 respectively, but Marchant (1968) has also found plants with *c.* 76 and *c.* 90 chromosomes. These plants he considers to be backcross progeny. The $2n = 90$ follows from a backcross to either parent (cytological evidence suggests to *S. alterniflora*) but $2n = 76$ can only be explained by a further backcross of the $2n = 90$ plants to *S. alterniflora* (Table 6.1).

Unfortunately, all attempts at simulating the origin of the *S.* x *townsendii* F_1 by crossing *S. maritima* and *S. alterniflora* under horticultural conditions have failed. Marchant suggests that transplantation experiments whereby *S. maritima* plants are introduced to natural *S. alterniflora* populations may be more successful.

One aspect of polyploidy is particularly puzzling and that is why polyploidy should be so common in plants and yet be rare in

Table 6.1

The origin of the chromosome races of the *Spartina* x *townsendii* aggregate.

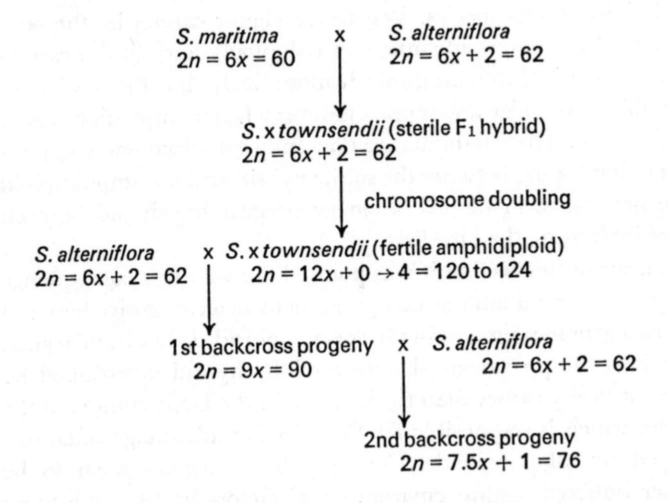

animals. Is there something about the biology of plants which enables them to make use of polyploidy whereas animals cannot? One suggestion is that the sex chromosome systems of animals would break down and indeed disruption of a sex chromosome mechanism can be demonstrated in the monoecious plant *Melandrium album* (= *Silene alba*) when the chromosome number is doubled. If this suggestion is correct then it is reasonable to expect that polyploidy may be common among hermaphrodite animals but in fact although it does occur in these groups it is far from being common.

What are the advantages of polyploidy?

It has often been noticed that polyploid plants are more luxuriant than their diploid parents and hence it has been argued that sheer vegetative vigour is the major advantage. Examination of cellular structure reveals that polyploids often have larger cells than their diploid relatives and obviously, because their nuclei contain twice as much chromosomal material, the nuclei will be larger.

It was originally thought that the success of the sterile hybrid of *S.* x *townsendii* was due to vegetative vigour but, if this were so, why do we not see more frequently in natural populations similar explosions of new species. Vegetative vigour cannot be the only or indeed the major advantage of polyploidy and, in the case of *S.* x *townsendii*, Marchant thinks it more likely that the new plant was able to colonize a vast open habitat where competition was at a minimum. These habitats are now fully colonized and competition is developing between the sterile hybrid and the amphidiploid The next stage in the evolutionary programme should be even more interesting than the first.

Earlier in this section it was pointed out that one consequence of chromosome doubling was permanent heterozygosity between the two genomes present in the species hybrid. It has been argued that it is the physiological nature of the hybrid determined by heterozygosity rather than the increase in the DNA content of the nuclei which is responsible for the selective advantage often displayed by allopolyploids. Certainly heterozygotes seem to be better buffered against environmental factors by having a more efficient homeostatic mechanism. This is borne out by the general observation of Stebbins that plant species which are tolerant of a wide range of conditions are often polyploids.

7

Problems of interpretation

Conclusions from statistical analysis

Statistical methods cannot be used to *prove* anything, because they are concerned with probabilities and not with the absolute. All we can say is that our results are more or less likely to agree with a hypothesis we have put forward. If the model matches the data with a reasonable degree of probability then the various parameters of the model may be estimated from the data. We can establish higher degrees of probability by multiple repetition of experiments, but beyond a certain point this becomes a waste of time. We have to accept a particular level of probability as being good enough.

But often one can be led astray. Consider for a moment intelligence and the ways of measuring it. As no two people agree on what intelligence is, it is very difficult to devise tests to estimate it. As has often been pointed out, all that an I.Q. test does is to test your ability to do the test; it is merely an interpretation by the man who devised the test that it measures intelligence. You sit the test, your attempt is marked and you can be ranked with your colleagues according to the marks you each gained. Does this mean that Bloggs, who scored 59 per cent, is more intelligent than Snooks, who managed only 55 per cent? Probably not we would say, but suppose on three other tests Bloggs scored 58, 62, and 56, while Snooks obtained 53, 54, and 51. Here is a consistent difference between the two examinees but what does this mean? All it means is that on this type of test Bloggs scores more marks than Snooks. It does not tell you necessarily that Bloggs is brighter than Snooks. The dilemma here is that we want to measure an entity which is indefinable and even though we design a test we still do not know whether the test measures what we want to measure. The same problem can arise when studying variation in plant populations.

Earlier we considered the effect of high concentrations of lead in the soil on the distribution of plants. Plants able to grow in culture solutions containing lead are said to be lead tolerant. But this is an interpretative assumption. Perhaps the presence of lead in the culture solution reduces the availability of some other ion and it is the ability of a mutant form to grow with a lowered threshold requirement of this ion which enables it to colonize lead-contaminated soil. We have, therefore, ambiguities, and these must be clarified or at least borne in mind before interpretations which make any sense can be suggested. Although it is possible to study the relationship between the phenotype and the environment we usually know next to nothing about the mechanisms by which the phenotype is produced. This is a major disadvantage of biometrical genetics as an analytical process. While on the one hand models which allow for dominance, additive effects, and non-allelic interaction go a long way towards explaining the gross gene-action, on the other hand they do not, and cannot, explain the underlying chemical basis of the phenotype. In plant and animal breeding work this often does not matter. One is interested in determining whether selected organisms have potential in a breeding programme and so knowledge of the gross genetical architecture is all that is required. We mentioned earlier the correlation between the frequency of cyanogenic *Trifolium repens* in Europe and January mean temperature. There was no justification for concluding that it was temperature which was controlling the distribution of the cyanogenic form until it could be shown that temperature had a direct, differential effect on the various phenotypes which occur. This Daday demonstrated later (1965), but until then he had a correlation from which it was not legitimate to draw conclusions about cause and effect, although it could be stated with some confidence that the frequency of cyanogenic plants did not cause the temperature difference!

When we obtain a significant correlation coefficient or a significant regression (see Appendix II for a discussion of these terms), we can then go on to calculate the equation of the straight line which best fits the data. Having done this we can estimate other points in the distribution. But making predictions from regression lines can be particularly hazardous. It is generally agreed that it is safe to predict only within the range of the original data. When extrapolations beyond the range are made, especially to determine y when $x = 0$, the conclusion drawn may be grossly inaccurate. When we calculate a regression equation from data we may be examining a short length of a quadratic or even higher order relationship. Extrapolation from part of a distribution when the true relationship is not linear throughout the whole distribution leads to all sorts of inaccurate predictions. Two errors of this nature in recent years were the predictions about the number of University places required during the period from 1970 to 1980, which was underestimated; and the birth-rate of coloured immigrants in the United Kingdom which was overestimated.

The question of scale

It has been pointed out repeatedly that infraspecific variation can be of many kinds, involving all sorts of characters and representing large or small responses to the habitat differences encountered by the species population. This question of the size of the response merits further discussion. The phenotypic differences observed range from one simple character, as in the case of the leaf spots in *Arum maculatum*, to a whole complex of characters such as those distinguishing the races of *Potentilla glandulosa*, which may be sufficiently distinct as a consequence to be recognized as separate subspecies. It is clear that the genetic mechanism behind these differences can correspondingly range from a single mendelian gene to a large number of genes spread over the whole genotype. This 'spread' may be more than a metaphorical expression, for the genes may be physically distributed on every chromosome of the complement.

Similar comparisons can be made of the habitats in which the plants are found. They may differ only in some minor soil constituent, or at the other extreme they may be on quite different soil types and in addition subjected to the contrasting climates of separate continents. The degree to which the habitats differ has a direct bearing on the variation found in the plants, because of the controlling effect of selection. A minor habitat difference will give rise to weak selection for certain character differences—a small difference in the calcium content of the soil, for instance, may have an effect which requires an elaborate physiological experiment to demonstrate. The single factor of the presence of a heavy metal may still only select a fairly simple response: that of tolerance to that particular metal, but the intensity of selection in this case is high and the effect big enough to be easily demonstrable. The next step up the scale takes us to the situation where not one but many character differences are selected by an equally complex set of environmental factors, of which the *Potentilla glandulosa* populations again afford a good example.

There are unexpected complications, however, to this idea that we can talk about every example in terms of its scale, and place it in a sequence of gradually increasing complexity. One of the best examples of the effects of selection on one or two major genes is that of cyanogenesis in *Trifolium* and *Lotus*. The degree of selection is either not very extreme, or it is balanced in some little understood way, because in most habitats the two forms coexist but there is a clear tendency for one to preponderate at low temperatures and the other at high. The complication is this: *Trifolium* was sampled over a very wide geographical and ecological range. Should we not have expected variation in many characters in response to what must have been highly complex selective forces? The answer is that such variation may well exist but that the author of the cyanogenesis study was not looking for it. In other words we cannot always expect someone interested in a particular character to take account of all the other possible characters at the same time. This is a most important point whenever we try to put together a comprehensive account of a species and the variation it shows. Inevitably such an account will have to be a synthesis of the work of many investigators, and draw on results which were often obtained for quite different purposes. The people investigating cyanogenesis were not setting out to map the overall pattern of variation in *Trifolium* and *Lotus* but to study one particular facet of its adaptation to different habitats. Even in Mooney and Billings' study of *Oxyria* (mountain sorrel) (1961), which was broader based in that they started with a pair of subspecies distinguished by a number of morphological characters the experimental work was confined to characters relevant to adaptation to climate. The chief habitat variables over their transect from the Rockies to Alaska were temperature and daylength and they found corresponding differences of temperature response in the physiological processes of the plants. It is quite certain, on the other hand, that the other features of the environment could not have remained uniform over the whole sampling area. There must have been, for example, considerable variation in the soils in which *Oxyria* was growing. It is no criticism of a very comprehensive and valuable piece of work to point out that adaptation to soil differences was not considered, although we may think it likely that they existed.

This concentration on particular characters is bound to influence any general discussion of scale. In very many cases we can only say that certain differences have been demonstrated, but that there may be many others. One of the distinctive features of modern taxonomic thought is the attempt to assess the relevance of all variation and to discuss the differences between two taxa in terms of the overall picture given by as wide an array of characters as possible. So far such ideas have been tried out mainly on morphological characters but there is a rapidly increasing body of biochemical and cytological information which is gradually being incorporated into taxonomy.

This question of scale arises most urgently when we try to orientate the type of variation discussed in this book in the wider context of evolution. The geneticist sometimes claims that any change of gene frequency is evolution of a sort, and while the logic of this can hardly be denied it is a somewhat unsatisfactory definition to anyone interested in, say, palaeontology. He sees the evidence of past evolution in the fossil record, and deals with big morphological changes and enormous lengths of time. The changes involved in the evolution of birds from reptiles can no doubt be thought of in terms of shifts of gene frequency but on such a massive scale that the geneticist cannot really offer a helpful account of the detailed processes involved. It is conceivable that one day we shall be able to make the direct comparison of the DNA sequences of members of distantly related groups, so that in one sense their genetical differences could be assessed. For the moment we can only make such comparisons by testing the two sets of chromosomes side by side in the same cell in the course of ordinary genetical analysis, and so we are restricted to organisms which can be intercrossed.

What is the connection between changes of gene frequency at the species and population levels and changes on a scale which the palaeontologist would recognize as evolution? What are the chances of one leading to the other? There is a great deal still to be learnt about this but in our present state of knowledge one key point seems clear and that is the importance of genetic isolation. As long as two populations with different genetic constitutions are potentially able to exchange genes the differences between them are not secure and may quite well be lost as a result of migration or geographical changes. Once a permanent barrier to crossing is established the two populations are independent and the effect of further selection will almost certainly be to increase their differences rather than reduce them.

The important link between genecology and evolution proper is that genetical diversity can in itself increase the chances of isolation arising. A species capable of a great deal of ecotypic variation will be able to spread into a wide variety of habitats and over a big geographical range. Such a species is much more likely to become subdivided into a number of isolated portions than is one with a smaller range. This seems to be the realistic approach to infraspecific variation in general and in particular to that portion of such variation which appears to offer adaptive advantages. It is not a question of a particular group of ecotypes being incipient species. There is nothing at all automatic about the change of status implied in that expression. Some ecotypes, and some portions of ecoclines, will become involved in isolating situations and may then progress to a higher status, but equally certainly many such ecotypes will not, and will merely be submerged again when the balance of selection and gene flow changes.

To some extent the striving for complete synthesis on the part of taxonomists and the analytical study of adaptation may be incompatible. Progress in the latter seems to depend on the isolation of small portions of the total variation—in other words single characters or readily comprehended groups of characters, which are simple enough to yield to experimental attack. Perhaps we shall gradually be able to tackle more ambitious projects, but there seems little hope at the moment of offering the taxonomist a functional explanation for more than a small fraction of the total variation with which he is faced. If he wishes to maintain that most of it is not adaptive we can only retort that evidence one way or the other is completely lacking. The chastening fact is that we are surrounded by variation in living things and that although we maintain the importance of natural selection in evolution in general terms, we have usually very little idea what the adaptive implication of any particular example may be.

Appendix I

Degrees of freedom

In Chapter 2 we introduced 'degrees of freedom' without defining in any way what this phrase meant. Although it was first used in relation to the analysis of discontinuous variation (χ^2), let us consider its meaning in relation to variance.

When we measure some character on, say, twenty individuals in a population we have 20 items of information; we have 20 degrees of freedom in choosing the individuals. To calculate the variance, however, we first have to calculate the mean and we use one degree of freedom to do so. Thus the variance we obtain has $20 - 1 = 19$ degrees of freedom.

There is another way we can look at the problem. If we have the mean of a sample of size 20, then 19 of the observations can be arbitrarily fixed but once this is done the twentieth is necessarily determined. We have 19 degrees of freedom in assigning values to our 20 observations. Thus the number of degrees of freedom is obtained as the number of classes which can be filled arbitrarily.

Furthermore, the sample mean is only an estimate of the true mean of the population. It can be shown arithmetically as well as algebraically that the variance calculated by using the sample mean is too small when the sum of squares is divided by n. Obviously the variance obtained will be larger when the sum of squares is divided by $n - 1$, but this variance is a closer approximation to the population variance, which we would obtain if we could use the true mean, than the one obtained by dividing the sum of squares by n. For a more rigorous mathematical discussion of these problems see, for example, Bulmer (1967).

Probability

We have to decide the level of probability at which we prefer to reject any hypothesis we may have put forward. When we carry out χ^2 tests we may be testing whether the observed results agree with those expected on some hypothesis. Alternatively we may be determining whether the ratios between two or more classes are the same as those between two or more other classes (Contingency χ^2). With the variance ratio test we are determining the probability that chance is responsible for the large mean square exceeding the smaller mean square. It is conventional to take the 5 per cent level of probability as the threshold. If the probability exceeds 5 per cent we accept the hypothesis, otherwise we reject it and have to think up a new one to explain the results. This presents two problems. Occasionally we could be led to a wrong decision and reject a hypothesis because every now and then extreme results will happen by chance. This is particularly noticeable when twenty or more similar experiments are carried out. It is not unreasonable to expect a few aberrant results and so we do not worry about these. Obviously if the aberrations are consistent then we should examine the cause more closely.

On the other hand the hypothesis we have suggested may not be the best interpretation of the results and yet the 5 per cent level of probability is not reached by the test applied. For example, two independent hypotheses may be equally efficient at explaining the data. Obviously both cannot be correct and we would have to choose one or the other by carrying out further experiments designed to distinguish between them.

Finally, it is necessary to distinguish the concept of statistical significance from that of biological importance. A statistical test may show that a difference exists but this does not necessarily mean that the difference we have demonstrated is of any importance biologically. The latter will usually depend upon the magnitude of this difference. Thus, having shown that a difference exists, some estimate of its size must be obtained before we can draw any useful conclusions. Almost certainly we will have to devise further experiments to verify these conclusions.

Appendix II

Calculations involving continuous variation

We will consider two situations of common occurrence in the study of plant populations.

(1) We sample two or more populations; do these populations differ with respect to the particular character we are examining?

(2). There may be an apparent relation between the size (say) of a character in different populations and some measurable factor of the environment; how can we tell whether this association is meaningful?

Analysis of variance

The best way to attack these problems is by dealing with actual situations, so consider the data in Table A1.

Table A1

The number of flowers per plant counted in two populations. Do the populations differ in the number of flowers produced ?

Population A	24	23	24	21	22	21	23	21	19	23
Population B	19	20	20	22	21	20	19	22	19	18

Here we have two populations in which the number of flowers produced by each of 10 plants has been counted. Do the populations differ in the production of flowers? By inspection it would seem that in population A the plants had a few more flowers than those in population B. Is this impression justifiable? The technique of Analysis of Variance helps us to answer these questions.

There are 20 observations, but because we use the mean of the sample to calculate the sum of squares (SS) we have used up one degree of freedom. There are therefore 19 degrees of freedom available for this analysis. Of these, one is associated with the difference between the populations while the other 18 are attached to the differences between the individuals within each population. The skeleton analysis becomes:

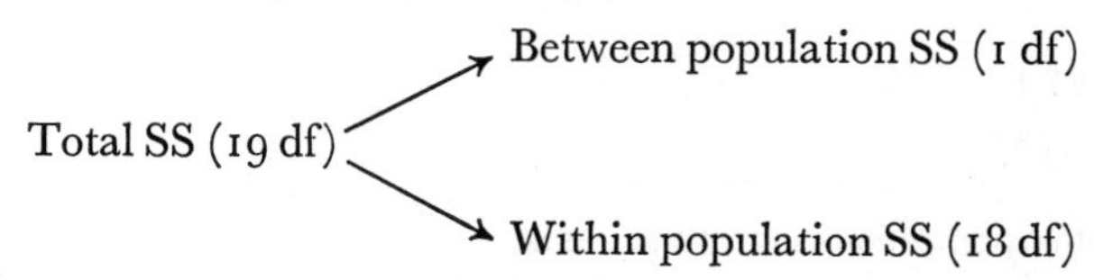

The job now is the calculation of the sums of squares associated with the two subdivisions given above. We will use the working mean method first as it does not require an elaborate calculating machine. A working mean is used in order to make the computation simpler and therefore less likely to arithmetic error.

Taking a working mean of 20, we construct Table A2.

Table A2

Population *A*	Deviation from working mean of 20 (x_A)	(Deviation)^2 $(x_A)^2$	Population *B*	Deviation from working mean (x_B)	(Deviation)^2 $(x_B)^2$
24	+4	16	19	−1	1
23	+3	9	20	0	0
24	+4	16	20	0	0
21	+1	1	22	+2	4
22	+2	4	21	+1	1
21	+1	1	20	0	0
23	+3	9	19	−1	1
21	+1	1	22	+2	4
19	−1	1	19	−1	1
23	+3	9	18	−2	4
221	$\sum x_A=21$	$\sum(x_A)^2=67$	200	$\sum x_B=0$	$\sum(x_B)^2=16$
	$n_A=10$ plants			$n_B=10$ plants	

$\bar{x}_A = \frac{\sum x_A}{n_A} = \frac{21}{10} = 2.1$ $\quad\therefore$ true mean = 20.0 + 2.1 = 22.1

$\bar{x}_B = 0$ $\quad\therefore$ true mean = 20.0

The total sum of squares is

$$\sum(x_A)^2 + \sum(x_B)^2 - \frac{(\sum x_A + \sum x_B)^2}{n_A + n_B}$$

$$= 67 + \quad 16 - \quad \frac{(21+0)^2}{20}$$

$$= 60.95$$

The within populations sum of squares is

$$\sum(x_A)^2 - \frac{(\sum x_A)^2}{n_A} + \sum(x_B)^2 - \frac{(\sum x_B)^2}{n_B}$$

$$= 67 - \quad 44.1 + \quad 16 - \quad 0$$

$$= 38.90$$

By subtraction, the between population sum of squares is

$$60.95 - 38.90 = 22.05$$

The latter can be checked by calculation from

$$\frac{(\sum x_A)^2}{n_A} + \frac{(\sum x_B)^2}{n_B} - \frac{(\sum x_A + \sum x_B)^2}{n_A + n_B} = 44.1 + 0 - 22.05 = 22.05$$

The analysis of variance can now be set out as in Table A3.

Table A3

Analysis of variance of flower number.

Item	Degrees of freedom	Sum of squares	Mean square (variance)	Variance ratio	Probability
Between populations	1	22.05	22.05	10.21	$0.01 > P > 0.001$
Within populations	18	38.90	2.16		
Total	19	60.95			

We have partitioned the sum of squares of 60.95 (which gives rise to an overall variance of $60.95/19 = 3.21$) according to two components. In this case, a sum of squares of 22.05 for one degree of freedom is due to the difference between the populations while the rest, 38.90 for 18 degrees of freedom, reflects the variation between the various plants within each population.

The variance ratio is the larger mean square (variance) divided by the smaller. This has attached, in this case, 1 and 18 degrees of freedom. When our calculated value is compared with the values in tables of variance ratio it is seen that our value is larger than that for 1 and 18 degrees of freedom in the 0.01 table, but smaller than the corresponding value in the 0.001 table. Thus the probability that the two variances are the same lies between 1 in 100 and 1 in 1000, i.e. they are significantly different. We would conclude that, although the numbers look very similar at first sight, there is a highly significant difference between the two populations with respect to the number of flowers produced by the plants.

The alternative technique, which is only to be recommended if a desk calculator or a computer is available, is shown in Table A4. The construction of the Analysis of Variance table is exactly the same as in Table A3.

In the first method we used a working mean of 20 while for the second the working mean was zero. We obtained the same figures on each occasion. We would, of course, have obtained exactly the same result if we had used the true sample mean of $421/20 = 21.05$ only the calculations would have been more tedious and consequently more liable to error. What this procedure shows is that it does not matter what sample mean is chosen, the variance obtained is a measure of the spread about the true mean. This can be proved algebraically.

There is an alternative test of significance which can be used here, and that is the t test. It is most convenient when we want to test pairs of samples and in this case we test for a significant difference between the two means.

i.e. $d = \bar{x}_A - \bar{x}_B$

but because $t = d/s_d$ we need to calculate s_d first

Table A4

Population A x_A	Square $(x_A)^2$	Population B x_B	Square $(x_B)^2$
24	576	19	361
23	529	20	400
24	576	20	400
21	441	22	484
22	484	21	441
21	441	20	400
23	529	19	361
21	441	22	484
19	361	19	361
23	529	18	324
$\sum x_A = 221$	$\sum (x_A)^2 = 4907$	$\sum x_B = 200$	$\sum (x_B)^2 = 4016$
$n_A = 10$ plants		$n_B = 10$ plants	

$$\text{Total SS} = \sum (x-\bar{x})^2 = 4907 + 4106 - \frac{(421)^2}{20} = 60.95$$

$$\text{Between populations SS} = \frac{221^2}{10} + \frac{200^2}{10} + \frac{(421)^2}{20} = 22.05$$

By subtraction the within populations SS = 60.95 − 22.05 = 38.90

(To check, the calculation is $\sum (x_A)^2 - \frac{(\sum x_A)^2}{n_A} + \sum (x_B)^2 - \frac{(\sum x_B)^2}{n_B} =$ 38·90).

We have already calculated $s_{\bar{x}}^2$ as 2.16 (within population mean square). It can be shown that $s_{\bar{d}}^2 = s_{\bar{x}}^2\ (\frac{1}{10} + \frac{1}{10})$ (there are two groups of ten numbers used to calculate $s_{\bar{x}}^2$)

thus $\quad s_{\bar{d}} = 2.16 \times 0.2 = 0.4322$

and $\quad s_{\bar{d}} = 0.6574$

hence $\quad t_{18} = d/s_{\bar{d}} = (22.1 - 20.0)/0.6574 = 3{\cdot}194$

From the tables of t it can be seen that this value of t for 18 degrees of freedom corresponds to values lying between a probability of

0.01 and 0.001, i.e. $0.01 > P > 0.001$. The difference between the mean flower number in the two populations is significant. Note that the value of $t = 3.194$ is the square root of the variance ratio of 10.21 for 1 and 18 degrees of freedom calculated earlier.

It is still possible to use the t test when more than two samples are being compared, but it is so much easier to use the analysis of variance that the latter method is recommended. Note at this stage that all we can conclude is that there are differences between populations A and B with respect to the number of flowers plants produce. More work is necessary to establish the nature of, and the reasons for, this difference—see Chapter 2, page 28.

As we have said previously, we suggest that the reader refers to the books listed at the end for the details of two- and three-factor analysis of variance and the analysis of field experiments involving more elaborate designs.

Regression analysis

We are now concerned with whether or not two variables are interrelated.

The only new item to be calculated is the sum of products of x and y which is written as SP (xy). The items we will need are (n = number of pairs of observations).

$$SS(x) = \sum_{i=1}^{n} \left(x_i - \bar{x}\right)^2 = \sum_{i=1}^{n} x_i^2 - \frac{\left(\sum_{i=1}^{n} x_i\right)^2}{n}$$

$$SS(y) = \sum_{i=1}^{n} y^2 - \frac{\left(\sum_{i=1}^{n} y_i\right)^2}{n}$$

$$SP(xy) = \sum_{i=1}^{n} \left(x_i - \bar{x}\right)\left(y_i - \bar{y}\right)$$

The last, for calculation is transformed to

$$SP(xy) = \sum_{i=1}^{n} x_i y_i - \frac{\left(\sum_{i=1}^{n} x_i\right)\left(\sum_{i=1}^{n} y_i\right)}{n}$$

with the obvious analogy with the easier formula for calculating sum of squares. $SP(xy)$ divided by $n - 1$ is called the covariance of x and y.

Now for an example. Consider the data in Table A5.

Table A5

***Festuca ovina*. The relation between the measured tolerance of plants and the lead content of the soil in which they were growing.**

Log_{10} of lead in soil (measured in parts per million) x	Tolerance level y	
1.77	17	$SS(x) = 5.794$
1.93	9	$SS(y) = 1634.000$
2.20	20	$SP(xy) = 90.680$
2.70	32	$\bar{x} = 2.954$
3.05	36	$y = 32.000$
3.25	30	$b = 15.651$ *
3.40	43	
3.42	35	(All these values were obtained without
3.48	47	using a working mean.)
4.34	51	*See text for explanation.

(For the purpose of illustration some of the data presented in Figure 4.2 has been tabulated here.)

In this example we are relating the mean tolerance of lead with a component of the environment, namely the lead content of the soil. Because it will remain the same at any one place whether the plants are there or not we regard the lead content of the soil as an independent variable, and the mean tolerance of the plants as a dependent variable. The total variation of the dependent variable is measured as $SS(y)$. Of this sum of squares the part which is accounted for by the independent variable is $\frac{(SP(xy))^2}{SS(x)}$ for one degree of freedom. We can obtain the residual sum of squares by subtraction and so the Analysis of Variance becomes:

Table A6

Analysis of variance of regression.

Item	df	SS	MS	VR
Regression	1	1419.193	1419.193	52.85
Residual	8	214.807	26.851	
Total		1634.000		

The variance ratio of 52.85 for 1 and 8 degrees of freedom corresponds with a probability of less than 0.001 and so we can conclude that the variance due to regression is significantly larger than the residual.

This regression analysis is essentially an extension of the co-ordinate geometry technique of finding the equation of the straight line which best fits a series of points on a graph. Taking the equation $y = mx + c$ we can estimate the values of m and c from the equation $y = \hat{b}(x - \bar{x}) + a$, where $a = \bar{y}$ and $\hat{b} = \frac{SP(xy)}{SS(x)}$.

Here $\bar{x} = 2.954$,

$\bar{y} = 32.000$

while $\hat{b} = 15.651$

so

$$y = 15.651\,(x - 2.954) + 32.000$$
$$= 15.651\,x - 14.233$$

Thus in the equation $y = mx + c$, $m = 15.651$ and $c = -14.233$. (One note of warning. If you use the working mean technique you must remember to add the working means to the estimated means before substituting in the equation $y = \hat{b}(x - \bar{x}) + a$.)

The regression line can be used to make predictions about the tolerance of plants growing in soils containing certain concentrations of lead. Suppose we find a soil with 100 parts per million of lead, what is the expected tolerance of the plants growing in that soil. Firstly we convert the concentration to logarithms to base 10. Thus 100 becomes 2.000. Substituting in the regression equation:

$$y = 15.651 \times 2.000 - 14.232$$
$$\therefore\ y = 17.070$$

The expected mean lead tolerance would be 17.070 and departures from this expected value can be tested using χ^2.

Frequency data

In other cases the data generated will be in terms of the frequency of various forms and this is especially true when we examine discontinuous variation. For example, we may determine the percentage of the plants in a population showing a particular character. In Table A7 there is a clear relation between the frequency of the character of cyanogenesis and altitude.

Table A7

Trifolium repens. Relation between altitude and the frequency of plants containing cyanogenic glucosides. (Data of Daday, *Heredity*, 1954, 8, 377).

	Locality					
	Lausanne	Naters	Fiesch	Alpage de Crausey	Kreuy	Gross Glockner
Height above sea level in feet	1903	2296	3510	4593	5577	6398
Per cent plants containing cyanogenic glucosides	85.87	81.00	50.91	13.33	8.51	0.00
Number of plants	184	100	100	90	94	99

The technique just described for regression analysis is adequate in very many cases and may be used with confidence. In a few situations it can, however, be misleading and what follows is a guide to more precise methods which give more reliable results.

The need for more elaborate methods lies in some of the rather peculiar properties of frequency data. Firstly, we have not taken into account the size of each sample. A large sample will obviously give more information than a small one. Indeed, when we consider the sample variance of frequency data it is found to be $\frac{pq}{n}$ (binomial variance), where p is the frequency of one form, q the frequency of the other ($p + q = 1$), and n is the sample size. Hence the larger n becomes the smaller is the variance and so the better the estimate of p.

Secondly, the variance $\frac{pq}{n}$ is dependent not only upon n but also upon p. Thus the variance itself varies according to the frequency of a particular form in a population and so frequency data are heterogeneous with respect to sampling variance. The techniques we have used up to now assume a constant variance in different samples, a requirement obviously not met by frequency data.

The first difficulty is easily overcome by ensuring that the sample sizes are approximately the same. Even the total of 184 plants at 1903 feet (in Table A7) is not so different from the others that it has a marked effect on the analysis, but it would certainly be better if all the samples were approximately 100.

The second difficulty was resolved many years ago by R. A. Fisher. He devised a method for transforming the data from percentages to angles measured in degrees. The sampling variance when this is done is $821/n$ for all values of p expressed as a percentage. And, of course, where n is the same from sample to sample the variance is also the same.

The actual transformation is called the arcsine transformation. $\theta = \sin^{-1} \sqrt{p}$ (also written as $\sin \theta = \sqrt{p}$). Fortunately there are tables for solving these ugly looking equations and one of them is included at the end of this book. All we have to do is to look up our particular value of p per cent, using the row and column margins and read off the value of θ from the body of the table. θ is used for the rest of the analysis. The tables can be used backwards for determining values of p per cent from given values of θ.

There is another minor problem which may arise with frequency data, an example occurring in the data presented in Table A7. When values of 0 or 100 per cent are included in the data these analyses can give misleading results and the standard procedure consists of replacing these two values by $\frac{0.25}{n} \times 100$ per cent and $\frac{n-0.25}{n} \times 100$ per cent respectively.

Because it will reduce the chance of arithmetic error when adding, it is sensible to turn the table round for the analysis, as in Table A8:

Table A8

Data of Table A7 transformed and rearranged.

Height in feet x	$p\%$	θ y		
1903	85.87	67.92	SS(x)	= 16 117 318.834
2296	81.00	64.16	SS(y)	= 3574.145
3510	50.91	45.52	SP(xy)	= −237 840.483
4593	13.33	21.41	$\bar{x}$	= 4046.167
5577	8.51	16.96	$\bar{y}$	= 36.473
6398	0.25*	2.87	$\hat{b}$	= −0.014 76

$\sin\theta = \sqrt{p}$ transformation used.

*Estimated as $\frac{0.25}{99} \times 100$ per cent (see text).

In this example we are relating the frequency of the plant character with altitude and because it will remain the same at any one place whether the plants are there or not, we take height above sea level as the independent variable, and frequency as the dependent variable. As before, the total variation of the dependent variable is measured as $SS(y)$, and so the analysis of variance becomes:

Table A9

Analysis of variance of regression.

Item	d.f.	SS	MS	VR
Regression	1	3590.771	3509.771	218.09
Residual	4	64.374	16.094	

The variance ratio of 218.14 for 1 and 4 degrees of freedom corresponds with a probability of less than 0.001 and hence the variance due to regression is significant.

The calculation of the regression line becomes:

$$y = 36.473 - 0.014\,76\,(x - 4046.167)$$
$$= 96.194 - 0.014\,76\,x$$

This regression equation can now be used to make predictions about the frequency of plants containing the cyanogenic glucoside at certain heights. For example, at 4000 feet the frequency would be

$$y = -0.014\,76 \times 4000 + 96.19$$
$$= -59.04 + 96.9$$
$$= 37.15 \text{ in angles measured in degrees.}$$

The expected frequency is 36.47 per cent.

Departures from the expected value, using the actual numbers observed and expected, NOT the frequencies, can be tested using χ^2.

A useful exercise is to calculate the equation of the regression line without using the angular transformation and then estimate p per cent at the heights given in Table A7. Follow this up by using the equation obtained from the transformed data—remember that the y you calculate on the second occasion needs to be transformed back from angles to percentages—and then satisfy yourself that the second method gives a better fit with the data.

When the analysis involves frequency data we can use the sampling error as a means of checking on the residual variance in the analysis of variance. We can show this best by using the example already described. The sampling variance of the transformed data is $\frac{821}{n}$ and this value can be used as the basic error of the whole analysis. Where the size of each sample is the same there is no problem in calculating $\frac{821}{n}$, but when the sample size varies, as it does in Table A7, the value of n which we use must be the harmonic mean of all the sample sizes.
In this case it is

$$\frac{6}{\left(\frac{1}{184}+\frac{1}{100}+\frac{1}{100}+\frac{1}{90}+\frac{1}{94}+\frac{1}{99}\right)} = 104.7391$$

where the 6 represents the number of samples taken.
The sampling variance becomes

$$\frac{821}{104.7391} = 7.839$$

We can now compare the residual variance with the sampling variance in the usual way. The variance ratio we obtain is

$$\frac{16.094}{7.839} = 2.05$$

for 1 and ∞ degrees of freedom. The probability is between 0.20 and 0.10 and so we conclude that the residual variance is not significantly larger than the sampling (error) variance. The interpretation here is that there is no evidence that the relationship between the frequency of the cyanogenic plants and altitude is not linear. A significant variance ratio would indicate that the relationship is quartic, exponential, logarithmic, etc.

The correlation coefficient

Correlation is another way of determining the relationship between two variables but it is less useful than regression analysis because only the latter can be used to estimate values of y for given values of x. In some cases it is not possible to distinguish between the dependent and independent variables, as for example in the length and the breadth of a leaf. In these situations the correlation coefficient is most useful. The interrelationship between correlation and regression becomes clearer on examination of the formula used to calculate the correlation coefficient. While calculating the regression sum of squares we computed all the necessary components for determining the correlation coefficient r. In our previous terminology

$$r = \frac{\sum_{i=1}^{n}\left(x_i - \bar{x}\right)\left(y_i - \bar{y}\right)}{\sqrt{\sum_{i=1}^{n}\left(x_i - \bar{x}\right)^2 \sum_{i=1}^{n}\left(y - \bar{y}\right)^2}} = \frac{\mathrm{SP}\left(xy\right)}{\sqrt{\mathrm{SS}\left(x\right) \times \mathrm{SS}\left(y\right)}}$$

for degrees of freedom two less than the number of pairs of observations. If there is no correlation then r tends towards zero but as the correlation increases, r approaches unity. Strictly the correlation coefficient is the covariance of x and y divided by the square root of the product of the variances of x and y but because $(n - 1)$ appears both in the numerator and denominator the formula given above is more useful for practical purposes.

Extensions

1. The relationship between two variables may not be linear, in which case the analysis given above is not suitable.
2. It may be that a variable is dependent upon two other variables which are themselves interrelated. By use of multiple regression analysis it is often possible to determine which of these variables has the greater effect. Reference to the books listed at the end is suggested.

Appendix III Statistical tables

Tables 1–5 are taken from Tables III, IV, V, VII and X of Fisher and Yates: *Statistical Tables for Biological, Agricultural and Medical Research*, published by Oliver Boyd Ltd., Edinburgh, and by permission of the authors and publishers.

Table 1 Distribution of *t*

N	Probability										
	0.90	0.80	0.70	0.50	0.30	0.20	0.10	0.05	0.02	0.01	0.001
1	0.16	0.33	0.51	1.00	1.96	3.08	6.31	12.71	31.82	63.66	636.62
2	0.14	0.29	0.45	0.82	1.39	1.89	2.92	4.30	6.97	9.93	31.60
3	0.14	0.28	0.42	0.77	1.25	1.64	2.35	3.18	4.54	5.84	12.94
4	0.13	0.27	0.41	0.74	1.19	1.53	2.13	2.78	3.75	4.60	8.61
5	0.13	0.27	0.41	0.73	1.16	1.48	2.02	2.57	3.37	4.03	6.86
6	0.13	0.27	0.40	0.72	1.13	1.44	1.94	2.45	3.14	3.71	5.96
7	0.13	0.26	0.40	0.71	1.12	1.42	1.90	2.37	3.00	3.50	5.41
8	0.13	0.26	0.40	0.71	1.11	1.40	1.86	2.31	2.90	3.36	5.04
9	0.13	0.26	0.40	0.70	1.10	1.38	1.83	2.26	2.82	3.25	4.78
10	0.13	0.26	0.40	0.70	1.09	1.37	1.81	2.23	2.76	3.17	4.59
11	0.13	0.26	0.40	0.70	1.09	1.36	1.80	2.20	2.72	3.11	4.44
12	0.13	0.26	0.40	0.70	1.08	1.36	1.78	2.18	2.68	3.06	4.32
13	0.13	0.26	0.39	0.69	1.08	1.35	1.77	2.16	2.65	3.01	4.22
14	0.13	0.26	0.39	0.69	1.08	1.35	1.76	2.15	2.62	2.98	4.14
15	0.13	0.26	0.39	0.69	1.07	1.34	1.75	2.13	2.60	2.95	4.07
16	0.13	0.26	0.39	0.69	1.07	1.34	1.75	2.12	2.58	2.92	4.02
17	0.13	0.26	0.39	0.69	1.07	1.33	1.74	2.11	2.57	2.90	3.97
18	0.13	0.26	0.39	0.69	1.07	1.33	1.73	2.10	2.55	2.88	3.92
19	0.13	0.26	0.39	0.69	1.07	1.33	1.73	2.09	2.54	2.86	3.88
20	0.13	0.26	0.39	0.69	1.06	1.33	1.73	2.09	2.53	2.85	3.85
22	0.13	0.26	0.39	0.69	1.06	1.32	1.72	2.07	2.51	2.82	3.79
24	0.13	0.26	0.39	0.69	1.06	1.32	1.71	2.06	2.49	2.80	3.75
26	0.13	0.26	0.39	0.68	1.06	1.32	1.71	2.06	2.48	2.78	3.71
28	0.13	0.26	0.39	0.68	1.06	1.31	1.70	2.05	2.47	2.76	3.67
30	0.13	0.26	0.39	0.68	1.06	1.31	1.70	2.04	2.46	2.75	3.65

Table 2 Distribution of χ^2

N	Probability 0.90	0.80	0.70	0.50	0.30	0.20	0.10	0.05	0.02	0.01	0.001
1	0.016	0.064	0.15	0.46	1.07	1.64	2.71	3.84	5.41	6.64	10.83
2	0.21	0.45	0.71	1.39	2.41	3.22	4.61	5.99	7.82	9.21	13.82
3	0.58	1.01	1.42	2.37	3.67	4.64	6.25	7.82	9.84	11.34	16.27
4	1.06	1.65	2.20	3.36	4.88	5.99	7.78	9.49	11.67	13.28	18.47
5	1.61	2.34	3.00	4.35	6.06	7.29	9.24	11.07	13.39	15.09	20.52
6	2.20	3.07	3.83	5.35	7.23	8.56	10.65	12.59	15.03	16.81	22.46
7	2.83	3.82	4.67	6.35	8.38	9.80	12.02	14.07	16.62	18.48	24.32
8	3.49	4.59	5.53	7.34	9.52	11.03	13.36	15.51	18.17	20.09	26.13
9	4.17	5.38	6.39	8.34	10.66	12.24	14.68	16.92	19.68	21.67	27.88
10	4.87	6.18	7.27	9.34	11.78	13.44	15.99	18.31	21.16	23.21	29.59
11	5.58	6.99	8.15	10.34	12.90	14.63	17.28	19.68	22.62	24.73	31.26
12	6.30	7.81	9.03	11.34	14.01	15.81	18.55	21.03	24.05	26.22	32.91
13	7.04	8.63	9.93	12.34	15.12	16.99	19.81	22.36	25.47	27.69	34.53
14	7.79	9.47	10.82	13.34	16.22	18.15	21.06	23.69	26.87	29.14	36.12
15	8.55	10.31	11.72	14.34	17.32	19.31	22.31	25.00	28.26	30.58	37.70
16	9.31	11.15	12.62	15.34	18.42	20.47	23.54	26.30	29.63	32.00	39.25
17	10.09	12.00	13.53	16.34	19.51	21.62	24.77	27.59	31.00	33.41	40.79
18	10.87	12.86	14.44	17.34	20.60	22.76	25.99	28.87	32.35	34.81	42.31
19	11.65	13.72	15.35	18.34	21.69	23.90	27.20	30.14	33.69	36.19	43.82
20	12.44	14.58	16.27	19.34	22.78	25.04	28.41	31.41	35.02	37.57	45.32
22	14.04	16.31	18.10	21.34	24.94	27.30	30.81	33.92	37.66	40.29	48.27
24	15.66	18.06	19.94	23.34	27.10	29.55	33.20	36.42	40.27	42.98	51.18
26	17.29	19.82	21.79	25.34	29.25	31.80	35.56	38.89	42.86	45.64	54.05
28	18.94	21.59	23.65	27.34	31.39	34.03	37.92	41.34	45.42	48.28	56.89
30	20.60	23.36	25.51	29.34	33.53	36.25	40.26	43.77	47.96	50.89	59.70

Table 3 Variance ratio

(i) 0.20 Probability Point

N_2 \ N_1	1	2	3	4	5	6	12	24	∞
1	9.5	12.0	13.1	13.7	14.0	14.3	14.9	15.2	15.6
2	3.6	4.0	4.2	4.2	4.3	4.3	4.4	4.4	4.5
3	2.7	2.9	2.9	3.0	3.0	3.0	3.0	3.0	3.0
4	2.4	2.5	2.5	2.5	2.5	2.5	2.5	2.4	2.4
5	2.2	2.3	2.3	2.2	2.2	2.2	2.2	2.2	2.1
6	2.1	2.1	2.1	2.1	2.1	2.1	2.0	2.0	2.0
7	2.0	2.0	2.0	2.0	2.0	2.0	1.9	1.9	1.8
8	2.0	2.0	2.0	1.9	1.9	1.9	1.8	1.8	1.7
9	1.9	1.9	1.9	1.9	1.9	1.8	1.8	1.7	1.7
10	1.9	1.9	1.9	1.8	1.8	1.8	1.7	1.7	1.6
11	1.9	1.9	1.8	1.8	1.8	1.8	1.7	1.6	1.6
12	1.8	1.8	1.8	1.8	1.7	1.7	1.7	1.6	1.5
13	1.8	1.8	1.8	1.8	1.7	1.7	1.6	1.6	1.5
14	1.8	1.8	1.8	1.7	1.7	1.7	1.6	1.6	1.5
15	1.8	1.8	1.8	1.7	1.7	1.7	1.6	1.5	1.5
16	1.8	1.8	1.7	1.7	1.7	1.6	1.6	1.5	1.4
17	1.8	1.8	1.7	1.7	1.7	1.6	1.6	1.5	1.4
18	1.8	1.8	1.7	1.7	1.6	1.6	1.5	1.5	1.4
19	1.8	1.8	1.7	1.7	1.6	1.6	1.5	1.5	1.4
20	1.8	1.8	1.7	1.7	1.6	1.6	1.5	1.5	1.4
22	1.8	1.7	1.7	1.6	1.6	1.6	1.5	1.4	1.4
24	1.7	1.7	1.7	1.6	1.6	1.6	1.5	1.4	1.3
26	1.7	1.7	1.7	1.6	1.6	1.6	1.5	1.4	1.3
28	1.7	1.7	1.7	1.6	1.6	1.6	1.5	1.4	1.3
30	1.7	1.7	1.6	1.6	1.6	1.5	1.5	1.4	1.3
60	1.7	1.7	1.6	1.6	1.5	1.5	1.4	1.3	1.2
120	1.7	1.6	1.6	1.5	1.5	1.5	1.4	1.3	1.1
∞	1.6	1.6	1.6	1.5	1.5	1.4	1.3	1.2	1.0

Table 5—continued

(ii) 0.05 Probability Point

N_2 \ N_1	1	2	3	4	5	6	12	24	∞
1	161.4	199.5	215.7	224.6	230.2	234.0	243.9	249.0	254.3
2	18.5	19.0	19.2	19.3	19.3	19.3	19.4	19.5	19.5
3	10.1	9.6	9.3	9.1	9.0	8.9	8.7	8.6	8.5
4	7.7	6.9	6.6	6.4	6.3	6.2	5.9	5.8	5.6
5	6.6	5.8	5.4	5.2	5.1	5.0	4.7	4.5	4.4
6	6.0	5.1	4.8	4.5	4.4	4.3	4.0	3.8	3.7
7	5.6	4.7	4.4	4.1	4.0	3.9	3.6	3.4	3.2
8	5.3	4.5	4.1	3.8	3.7	3.6	3.3	3.1	2.9
9	5.1	4.3	3.9	3.6	3.5	3.4	3.1	2.9	2.7
10	5.0	4.1	3.7	3.5	3.3	3.2	2.9	2.7	2.5
11	4.8	4.0	3.6	3.4	3.2	3.1	2.8	2.6	2.4
12	4.8	3.9	3.5	3.3	3.1	3.0	2.7	2.5	2.3
13	4.7	3.8	3.4	3.2	3.0	2.9	2.6	2.4	2.2
14	4.6	3.7	3.3	3.1	3.0	2.9	2.5	2.3	2.1
15	4.5	3.7	3.3	3.1	2.9	2.8	2.5	2.3	2.1
16	4.5	3.6	3.2	3.0	2.9	2.7	2.4	2.2	2.0
17	4.5	3.6	3.2	3.0	2.8	2.7	2.4	2.2	2.0
18	4.4	3.6	3.2	2.9	2.8	2.7	2.3	2.1	1.9
19	4.4	3.5	3.1	2.9	2.7	2.6	2.3	2.1	1.9
20	4.4	3.5	3.1	2.9	2.7	2.6	2.3	2.1	1.8
22	4.3	3.4	3.1	2.8	2.7	2.6	2.2	2.0	1.8
24	4.3	3.4	3.0	2.8	2.6	2.5	2.2	2.0	1.7
26	4.2	3.4	3.0	2.7	2.6	2.5	2.2	2.0	1.7
28	4.2	3.3	3.0	2.7	2.6	2.4	2.1	1.9	1.7
30	4.2	3.3	2.9	2.7	2.5	2.4	2.1	1.9	1.6
60	4.0	3.2	2.8	2.5	2.5	2.3	1.9	1.7	1.4
120	3.9	3.1	2.7	2.5	2.3	2.2	1.8	1.6	1.3
∞	3.8	3.0	2.6	2.4	2.2	2.1	1.8	1.5	1.0

Table 3–continued

(iii) 0.01 Probability Point

N_2 \ N_1	1	2	3	4	5	6	12	24	∞
1	4,052	4,999	5,403	5,625	5,764	5,859	6,106	6,234	6,366
2	98.5	99.0	99.2	99.3	99.3	99.3	99.4	99.5	99.5
3	34.1	30.8	29.5	28.7	28.2	27.9	27.1	26.6	26.1
4	21.2	18.0	16.7	16.0	15.5	15.2	14.4	13.9	13.5
5	16.3	13.3	12.1	11.4	11.0	10.7	9.9	9.5	9.0
6	13.7	10.9	9.8	9.2	8.8	8.5	7.7	7.3	6.9
7	12.3	9.6	8.5	7.9	7.5	7.2	6.5	6.1	5.7
8	11.3	8.7	7.6	7.0	6.6	6.4	5.7	5.3	4.9
9	10.6	8.0	7.0	6.4	6.1	5.8	5.1	4.7	4.3
10	10.0	7.6	6.6	6.0	5.6	5.4	4.7	4.3	3.9
11	9.7	7.2	6.2	5.7	5.3	5.1	4.4	4.0	3.6
12	9.3	6.9	6.0	5.4	5.1	4.8	4.2	3.8	3.4
13	9.1	6.7	5.7	5.2	4.9	4.6	4.0	3.6	3.2
14	8.9	6.5	5.6	5.0	4.7	4.5	3.8	3.4	3.0
15	8.7	6.4	5.4	4.9	4.6	4.3	3.7	3.3	2.9
16	8.5	6.2	5.3	4.8	4.4	4.2	3.6	3.2	2.8
17	8.4	6.1	5.2	4.7	4.3	4.1	3.5	3.1	2.7
18	8.3	6.0	5.1	4.6	4.3	4.0	3.4	3.0	2.6
19	8.2	5.9	5.0	4.5	4.2	3.9	3.3	2.9	2.5
20	8.1	5.9	4.9	4.4	4.1	3.9	3.2	2.9	2.4
22	7.9	5.7	4.8	4.3	4.0	3.8	3.1	2.8	2.3
24	7.8	5.6	4.7	4.2	3.9	3.7	3.0	2.7	2.2
26	7.7	5.5	4.6	4.1	3.8	3.6	3.0	2.6	2.1
28	7.6	5.5	4.6	4.1	3.8	3.5	2.9	2.5	2.1
30	7.6	5.4	4.5	4.0	3.7	3.5	2.8	2.5	2.0
60	7.1	5.0	4.1	3.7	3.3	3.1	2.5	2.1	1.6
120	6.9	4.8	4.0	3.5	3.2	3.0	2.3	2.0	1.4
∞	6.6	4.6	3.8	3.3	3.0	2.8	2.2	1.8	1.0

Table 3–continued

(iv) 0.001 Probability Point

N_2 \ N_1	1	2	3	4	5	6	12	24	∞
1	405,284	500,000	540,379	562,500	576,405	585,937	610,667	623,497	636,619
2	998.5	999.0	999.2	999.2	999.3	999.3	999.4	999.5	999.5
3	167.5	148.5	141.1	137.1	134.6	132.8	128.3	125.9	123.5
4	74.1	61.3	56.2	53.4	51.7	50.5	47.4	45.8	44.1
5	47.0	36.6	33.2	31.1	29.8	28.8	26.4	25.1	23.8
6	35.5	27.0	23.7	21.9	20.8	20.0	18.0	16.9	15.8
7	29.2	21.7	18.8	17.2	16.2	15.5	13.7	12.7	11.7
8	25.4	18.5	15.8	14.4	13.5	12.9	11.2	10.3	9.3
9	22.9	16.4	13.9	12.6	11.7	11.1	9.6	8.7	7.8
10	21.0	14.9	12.6	11.3	10.5	9.9	8.5	7.6	6.8
11	19.7	13.8	11.6	10.4	9.6	9.1	7.6	6.9	6.0
12	18.6	13.0	10.8	9.6	8.9	8.4	7.0	6.3	5.4
13	17.8	12.3	10.2	9.1	8.4	7.9	6.5	5.8	5.0
14	17.1	11.8	9.7	8.6	7.9	7.4	6.1	5.4	4.6
15	16.6	11.3	9.3	8.3	7.6	7.1	5.8	5.1	4.3
16	16.1	11.0	9.0	7.9	7.3	6.8	5.6	4.9	4.1
17	15.7	10.7	8.7	7.7	7.0	6.6	5.3	4.6	3.9
18	15.4	10.4	8.5	7.5	6.8	6.4	5.1	4.5	3.7
19	15.1	10.2	8.3	7.3	6.6	6.2	5.0	4.3	3.5
20	14.8	10.0	8.1	7.1	6.5	6.0	4.8	4.2	3.4
22	14.4	9.6	7.8	6.8	6.2	5.8	4.6	3.9	3.2
24	14.0	9.3	7.6	6.6	6.0	5.6	4.4	3.7	3.0
26	13.7	9.1	7.4	6.4	5.8	5.4	4.2	3.6	2.8
28	13.5	8.9	7.2	6.3	5.7	5.2	4.1	3.5	2.7
30	13.3	8.8	7.1	6.1	5.5	5.1	4.0	3.4	2.6
60	12.0	7.8	6.2	5.3	4.8	4.4	3.3	2.7	1.9
120	11.4	7.3	5.8	5.0	4.4	4.0	3.0	2.4	1.6
∞	10.8	6.9	5.4	4.6	4.1	3.7	2.7	2.1	1.0

Table 4 Correlation coefficient

5 per cent and 1 per cent levels of significance

Degrees of Freedom	5%	1%	Degrees of Freedom	5%	1%
1	0.997	1.000	24	0.388	0.496
2	0.950	0.990	25	0.381	0.487
3	0.878	0.959	26	0.374	0.478
4	0.811	0.917	27	0.367	0.470
5	0.754	0.874	28	0.361	0.463
6	0.707	0.834	29	0.355	0.456
7	0.666	0.798	30	0.349	0.449
8	0.632	0.765	35	0.325	0.418
9	0.602	0.735	40	0.304	0.393
10	0.576	0.708	45	0.288	0.372
11	0.553	0.684	50	0.273	0.354
12	0.532	0.661	60	0.250	0.325
13	0.514	0.641	70	0.232	0.302
14	0.497	0.623	80	0.217	0.283
15	0.482	0.606	90	0.205	0.267
16	0.468	0.590	100	0.195	0.254
17	0.456	0.575			
18	0.444	0.561			
19	0.433	0.549			
20	0.423	0.537			
21	0.413	0.526			
22	0.404	0.515			
23	0.396	0.505			

Table 5 Angular transformation

p%	0.0	0.1	0.2	0.3	0.4	0.5	0.6	0.7	0.8	0.9
0	0.00	1.81	2.56	3.14	3.63	4.05	4.44	4.80	5.13	5.44
1	5.74	6.02	6.29	6.55	6.80	7.03	7.27	7.49	7.71	7.92
2	8.13	8.33	8.53	8.72	8.91	9.10	9.28	9.46	9.63	9.80
3	9.97	10.14	10.30	10.47	10.63	10.78	10.94	11.09	11.24	11.39
4	11.54	11.68	11.83	11.97	12.11	12.25	12.38	12.52	12.66	12.79
5	12.92	13.05	13.18	13.31	13.44	13.56	13.69	13.81	13.94	14.06
6	14.18	14.30	14.42	14.54	14.65	14.77	14.89	15.00	15.12	15.23
7	15.34	15.45	15.56	15.68	15.79	15.89	16.00	16.11	16.22	16.32
8	16 43	16.54	16.64	16.74	16.85	16.95	17.05	17.15	17.26	17 36
9	17.46	17.56	17.66	17.76	17.85	17.95	18.05	18.15	18.24	18.34
10	18.43	18.53	18.63	18.72	18.81	18.91	19.00	19.09	19.19	19.28
11	19.37	19.46	19.55	19.64	19.73	19.82	19.91	20.00	20.09	20.18
12	20.27	20.36	20.44	20.53	20.62	20.70	20.79	20.88	20.96	21.05
13	21.13	21.22	21.30	21.39	21.47	21.56	21.64	21.72	21.81	21.89
14	21.97	22.06	22.14	22.22	22.30	22.38	22.46	22.54	22.63	22.71
15	22.79	22.87	22.95	23.03	23.11	23.18	23.26	23.34	23.42	23.50
16	23.58	23.66	23.73	23.81	23.89	23.97	24.04	24.12	24.20	24.27
17	24.35	24.43	24.50	24.58	24.65	24.73	24.80	24.88	24.95	25.03
18	25.10	25.18	25.25	25.33	25.40	25.47	25.55	25.62	25.70	25.77
19	25.84	25.91	25.99	26.06	26.13	26.21	26.28	26.35	26.42	26.49
20	26.57	26.64	26.71	26.78	26.85	26.92	26.99	27.06	27.13	27.20
21	27.27	27.35	27.42	27.49	27.56	27.62	27.69	27.76	27.83	27.90
22	27.97	28.04	28.11	28.18	28.25	28.32	28.39	28.45	28.52	28.59
23	28.66	28.73	28.79	28.86	28.93	29.00	29.06	29.13	29.20	29.27
24	29.33	29.40	29.47	29.53	29.60	29.67	29.73	29.80	29.87	29.93
25	30.00	30.07	30.13	30.20	30.26	30.33	30.40	30.46	30.53	30.59
26	30.66	30.72	30.79	30.85	30.92	30.98	31.05	31.11	31.18	31.24
27	31.31	31.37	31.44	31.50	31.56	31.63	31.69	31.76	31.82	31.88
28	31.95	32.01	32.08	32.14	32.20	32.27	32.33	32.39	32.46	32.52
29	32.58	32.65	32.71	32.77	32.83	32.90	32.96	33.02	33.09	33.15
30	33.21	33.27	33.34	33.40	33.46	33.52	33.58	33.65	33.71	33.77
31	33.83	33.90	33.96	34.02	34.08	34.14	34.20	34.27	34.33	34.39
32	34.45	34.51	34.57	34.63	34.70	34.76	34.82	34.88	34.94	35.00
33	35.06	35.12	35.18	35.24	35.30	35.37	35.43	35.49	35.55	35.61

Table 5–continued

p%	0.0	0.1	0.2	0.3	0.4	0.5	0.6	0.7	0.8	0.9
34	35.67	35.73	35.79	35.85	35.91	35.97	36.03	36.09	36.15	36.21
35	36.27	36.33	36.39	36.45	36.51	36.57	36.63	36.69	36.75	36.81
36	36.87	36.93	36.99	37.05	37.11	37.17	37.23	37.29	37.35	37.41
37	37.46	37.52	37.58	37.64	37.70	37.76	37.82	37.88	37.94	38.00
38	38.06	38.12	38.17	38.23	38.29	38.35	38.41	38.47	38.53	38.59
39	38.65	38.70	38.76	38.82	38.88	38.94	39.00	39.06	39.11	39.17
40	39.23	39.29	39.35	39.41	39.47	39.52	39.58	39.64	39.70	39.76
41	39.82	39.87	39.93	39.99	40.05	40.11	40.16	40.22	40.28	40.34
42	40.40	40.45	40.51	40.57	40.63	40.69	40.74	40.80	40.86	40.92
43	40.98	41.03	41.09	41.15	41.21	41.27	41.32	41.38	41.44	41.50
44	41.55	41.61	41.67	41.73	41.78	41.84	41.90	41.96	42.02	42.07
45	42.13	42.19	42.25	42.30	42.36	42.42	42.48	42.53	42.59	42.65
46	42.71	42.76	42.82	42.88	42.94	42.99	43.05	43.11	43.17	43.22
47	43.28	43.34	43.39	43.45	43.51	43.57	43.62	43.68	43.74	43.80
48	43.85	43.91	43.97	44.03	44.08	44.14	44.20	44.26	44.31	44.37
49	44.43	44.48	44.54	44.60	44.66	44.71	44.77	44.83	44.89	44.94
50	45.00	45.06	45.11	45.17	45.23	45.29	45.34	45.40	45.46	45.52
51	45.57	45.63	45.69	45.74	45.80	45.86	45.92	45.97	46.03	46.09
52	46.15	46.20	46.26	46.32	46.38	46.43	46.49	46.55	46.61	46.66
53	46.72	46.78	46.83	46.89	46.95	47.01	47.06	47.12	47.18	47.24
54	47.29	47.35	47.41	47.47	47.52	47.58	47.64	47.70	47.75	47.81
55	47.87	47.93	47.98	48.04	48.10	48.16	48.22	48.27	48.33	48.39
56	48.45	48.50	48.56	48.62	48.68	48.73	48.79	48.85	48.91	48.97
57	49.02	49.08	49.14	49.20	49.26	49.31	49.37	49.43	49.49	49.55
58	49.60	49.66	49.72	49.78	49.84	49.89	49.95	50.01	50.07	50.13
59	50.18	50.24	50.30	50.36	50.42	50.48	50.53	50.59	50.65	50.71
60	50.77	50.83	50.89	50.94	51.00	51.06	51.12	51.18	51.24	51.30
61	51.35	51.41	51.47	51.53	51.59	51.65	51.71	51.77	51.83	51.88
62	51.94	52.00	52.06	52.12	52.18	52.24	52.30	52.36	52.42	52.48
63	52.54	52.59	52.65	52.71	52.77	52.83	52.89	52.95	53.01	53.07
64	53.13	53.19	53.25	53.31	53.37	53.43	53.49	53.55	53.61	53.67
65	53.73	53.79	53.85	53.91	53.97	54.03	54.09	54.15	54.21	54.27
66	54.33	54.39	54.45	54.51	54.57	54.63	54.70	54.76	54.82	54.88

p%	0.0	0.1	0.2	0.3	0.4	0.5	0.6	0.7	0.8	0.9
67	54.94	55.00	55.06	55.12	55.18	55.24	55.30	55.37	55.43	55.49
68	55.55	55.61	55.67	55.73	55.80	55.86	55.92	55.98	56.04	56.10
69	56.17	56.23	56.29	56.35	56.42	56.48	56.54	56.60	56.66	56.73
70	56.79	56.85	56.91	56.98	57.04	57.10	57.17	57.23	57.29	57.35
71	57.42	57.48	57.54	57.61	57.67	57.73	57.80	57.86	57.92	57.99
72	58.05	58.12	58.18	58.24	58.31	58.37	58.44	58.50	58.56	58.63
73	58.69	58.76	58.82	58.89	58.95	59.02	59.08	59.15	59.21	59.28
74	59.34	59.41	59.47	59.54	59.60	59.67	59.74	59.80	59.87	59.93
75	60.00	60.07	60.13	60.20	60.27	60.33	60.40	60.47	60.53	60.60
76	60.67	60.73	60.80	60.87	60.94	61.00	61.07	61.14	61.21	61.27
77	61.34	61.41	61.48	61.55	61.61	61.68	61.75	61.82	61.89	61.96
78	62.03	62.10	62.17	62.24	62.31	62.38	62.44	62.51	62.58	62.65
79	62.73	62.80	62.87	62.94	63.01	63.08	63.15	63.22	63.29	63.36
80	63.43	63.51	63.58	63.65	63.72	63.79	63.87	63.94	64.01	64.09
81	64.16	64.23	64.30	64.38	64.45	64.53	64.60	64.67	64.75	64.82
82	64.90	64.97	65.05	65.12	65.20	65.27	65.35	65.42	65.50	65.57
83	65.65	65.73	65.80	65.88	65.96	66.03	66.11	66.19	66.27	66.34
84	66.42	66.50	66.58	66.66	66.74	66.82	66.89	66.97	67.05	67.13
85	67.21	67.29	67.37	67.46	67.54	67.62	67.70	67.78	67.86	67.94
86	68.03	68.11	68.19	68.28	68.36	68.44	68.53	68.61	68.70	68.78
87	68.87	69.95	69.04	69.12	69.21	69.30	69.38	69.47	69.56	69.64
88	69.73	69.82	69.91	70.00	70.09	70.18	70.27	70.36	70.45	70.54
89	70.63	70.72	70.81	70.91	71.00	71.09	71.19	71.28	71.37	71.47
90	71.57	71.66	71.76	71.85	71.95	72.05	72.15	72.24	72.34	72.44
91	72.54	72.64	72.74	72.85	72.95	73.05	73.15	73.26	73.36	73.46
92	73.57	73.68	73.78	73.89	74.00	74.11	74.21	74.32	74.44	74.55
93	74.66	74.77	74.88	75.00	75.11	75.23	75.35	75.46	75.58	75.70
94	75.82	75.94	76.06	76.19	76.31	76.44	76.56	76.69	76.82	76.95
95	77.08	77.21	77.34	77.48	77.62	77.75	77.89	78.03	78.17	78.32
96	78.46	78.61	78.76	78.91	79.06	79.22	79.37	79.53	79.70	79.86
97	80.03	80.20	80.37	80.54	80.72	80.90	81.09	81.28	81.47	81.67
98	81.87	82.08	82.29	82.51	82.73	82.97	83.20	83.45	83.71	83.98
99	84.26	84.56	84.87	85.20	85.56	85.95	86.37	86.86	87.44	88.19

References and further reading

BARBER, H. N. (1955). 'Adaptive gene-substitutions in *Eucalyptus.*' *Evolution,* **9,** 1–14.

BARBER, H. N. (1965). 'Selection in natural populations'. *Heredity,* **20,** 551–72.

BISHOP, J. A. and KORN, M. E. (1969). 'Natural selection and cyanogenesis in white clover, *Trifolium repens*'. *Heredity,* **24,** 423–30.

BJÖRKMAN, O. and HOLMGREN, P. (1963) 'Adaptability of the photosynthetic apparatus to light intensity in ecotypes from exposed and shaded habitats.' *Physiologia Plantarum,* **16,** 889–914.

BODMER, W. F. (1960). 'The genetics of homostyly in populations of *Primula vulgaris*' *Phil. Trans. B,* **242,** 517–49.

BRADSHAW, A. D. (1965). 'The evolutionary significance of phenotypic plasticity.' *Advances in Genetics,* **13,** 115–55.

BRIGGS, D. and WALTERS, S. M. (1969). *Plant Variation and Evolution,* Weidenfeld & Nicolson.

BULMER, M. J. (1967). *Principles of Statistics,* Oliver & Boyd, 2nd Ed.

BUMPUS, H. C. (1899). 'The elimination of the unfit as illustrated by the introduced sparrow *Passer domesticus.*' Biol. Lect., Marine Biol. Labs. Woods Hole.

BÖCHER, T. W. (1949). 'Racial differences in *Prunella vulgaris* in relation to habitat and climate.' *New Phytologist,* **48,** 285–314.

CAMPBELL, R. C. (1967). *Statistics for Biologists,* Cambridge University Press.

CLAPHAM, A. R., TUTIN, T. G., and WARBURG, E. F. (1962). *Flora of the British Isles,* Cambridge University Press, 2nd Ed.,

CLATWORTHY, J. N. and HARPER, J. L. (1961). 'The Comparative biology of closely related species growing in the same area. V.' *J. exp. Bot.,* **13,** 307–24.

CLAUSEN, J. (1951). *Evolution of Plant Species,* Cornell University Press.

CLAUSEN, J., KECK, D. D., and HEISEY, W. M. (1940). *Experimental Studies on the Nature of Species I. Effect of varied environments on Western North American Plants,* Carnegie Institution of Washington. Publication No. 520.

COOK, L. M. (1971). *Coefficients of Natural Selection,* Hutchinson.

CRANE, M. B. and BROWN, A. C. (1937) 'Incompatibility in the sweet cherry *Prunus avium* L.' *J. Genetics*, **15**, 86–117.

CROSBY, J. L. (1959). 'Outcrossing in homostyle primroses.' *Heredity*, **13**, 127–31.

DADAY, H. (1954). 'Gene frequencies in wild populations of *Trifolium repens* L. I. Distribution by latitude.' *Heredity*, **8**, 61–78.

DADAY, H. (1965). 'Gene frequencies in wild populations of *Trifolium repens* L. IV. Mechanism of Natural Selection.' *Heredity*, **20**, 355–65.

DARWIN, C. R. (1859). *On the Origin of Species by means of Natural Selection*, Murray.

DAWSON, C. D. R. (1941). 'Tetrasomic inheritance in *Lotus corniculatus* L.' *J. Genetics*, **42**, 49–72.

DOWDESWELL, W. H. (1958). *The Mechanism of Evolution*, Heinemann Educational Books, 2nd Ed.

DOWRICK, V. P. J. (1956). 'Heterostyly and homostyly in *Primula obconica*' *Heredity*, **10**, 219–36.

EMERSON, S. (1939). 'A preliminary survey of the *Oenothera organensis* population.' *Genetics*, **24**, 524–37.

FLOR, H. H. (1956). 'The complementary genic systems in flax and flax rust.' *Advances in Genetics*, **8**, 29–54.

FORD, E. B. (1940). In *The New Systematics*, edited by J. S. Huxley, Clarendon Press.

FORD, E. B. (1971). *Ecological Genetics*, Chapman and Hall, 3rd Ed.

GRANT, V. (1963). *Origin of Adaptations*. Columbia University Press.

GRANT, V. (1965). 'Evidence for the selective origin of incompatibility barriers in the leafy-stemmed Gilias.' *Proc. Nat. Acad., Sci.*, **54**, 1567–71.

GREGOR, J. W. (1938). 'Experimental Taxonomy II. Initial population differentiation in *Plantago maritima* L. of Britian.' *New Phytologist*, **37**, 15–49.

GREIG-SMITH, P. (1964). *Quantitative Plant Ecology*, Butterworth, 2nd Ed.

HARBERD, D. J. (1961). 'Observations on population structure and longevity of *Festuca rubra* L.' *New Phytologist*, **60**, 184–206.

HARPER, J. L. (1961). 'Approaches to the study of plant competition' in *Mechanisms in biological competition*, edited by F. L. Milthorpe, Cambridge University Press.

HASKELL, G. M. (1953). 'Quantitative variation in subsexual *Rubus*.' *Heredity*, **7**, 409–18.

HASKELL, G. M. (1961). 'Genetics and the distribution of *British Rubi*.' *Genetica*, **32**, 118–32.

HESLOP-HARRISON, J. (1953). *New Concepts in Flowering Plant Taxonomy*, Heinemann Educational Books.

HESLOP-HARRISON, J. (1964). 'Forty years of genecology.' *Adv. Ecol. Res.*, **2**, 159–247.

HEWITT, E. J. (1966). *Sand and water culture methods.* Commonwealth Agricultural Bureaux, Tech. Publ. 22, 2nd Ed.

HUSKINS, C. L. (1931). 'Origin of *Spartina townsendii.*' *Nature,* **127,** 781; and *Genetica,* **12,** 531–8.

HUTCHINSON, T. C. (1967). 'Ecotype differentiation in *Teucrium scorodonia,* with respect to susceptibility to lime-induced chlorosis and to shade factors.' *New Phytologist,* **66,** 439–53.

HUXLEY, J. S. (1942). *Evolution: The Modern Synthesis,* Allen & Unwin.

JAIN, S. K. and BRADSHAW, A. D. (1966). 'Evolution in closely adjacent plant populations. I. The evidence and its theoretical analysis.' *Heredity,* **20,** 407–41.

JARVIS, P. G. (1964). 'The adaptability to light intensity of seedlings of *Quercus petraea* (Matt.) Liebl.' *J. Ecology,* **52,** 545–71.

JONES, D. A. (1966). 'On the polymorphism of cyanogenesis in *Lotus corniculatus* L. Selection by Animals.' *Canad. J. Genet. Cytol.,* **8,** 556–67.

JONES, D. A. (1970). 'On the polymorphism of cyanogenesis in *Lotus corniculatus* L. III. Some aspects of selection.' *Heredity,* **25,** 633–41.

JONES, M. E. (1971). 'The population genetics of *Arabidopsis thaliana.* III The effect of vernalization.' *Heredity,* **26,** (In press).

LAWRENCE, M. J. (1965). 'Variation in wild populations of *Papaver dubium.* I. Variation within Populations; Diallel Crosses.' *Heredity,* **20,** 183–204.

LEWIS, D. (1942). 'The evolution of sex in flowering plants.' *Biol. Reviews,* **17,** 46–67.

LEWIS, D. (1954). 'Comparative incompatibility in angiosperms and fungi' *Advances in Genetics,* **6,** 235–85.

LEWIS, K. R. and JOHN B. (1963). *Chromosome Marker,* Churchill.

MCMILLAN, C. (1965). 'Ecotypic differentiation within four North American prairie grasses. II. Behavioural variation within transplanted community fractions.' *Amer. J. Bot.,* **52,** 55–65.

MCNAUGHTON, S. J. (1965). 'Differential enzymatic activity in ecological races of *Typha latifolia* L.' *Science,* **150,** 1829–30.

MCNEILLY, T. (1968). 'Evolution in closely adjacent plant populations. III. *Agrostis tenuis* on a small coppermine.' *Heredity,* **23,** 99–108.

MCVEAN, D. N. (1953). 'Regional variation of *Alnus glutinosa* (L.) Gaerta. in Britain.' *Watsonia,* **3,** 26–32.

MARCHANT, C. J. (1968). 'Evolution in *Spartina* (Gramineae) II. Chromosomes, basic relationships and the problem of the *S.* x *townsendii* agg.' *J. Linn. Soc. (Bot.),* **60,** 381–409.

MARSDEN-JONES, E. M. and TURRILL, W. B. (1938). 'Transplant experiments of the British Ecological Society at Potterne, Wilts. Summary of results, 1928–1937.' *J. Ecology,* **26,** 380–9.

MATHER, K. (1940). 'Outbreeding and the separation of the sexes.' *Nature,* **145,** 484–6.

MATHER, K. (1951). *The Elements of Biometry*, Methuen.

MAYER, A. M. and POLJAKOFF-MAYBER, A. (1963). *The Germination of Seeds*, Pergamon.

MAYNARD-SMITH, J. (1966). *The Theory of Evolution*, Penguin Books, 2nd Ed.

MAYR, E. (1954). In *Evolution as a Process*, edited by J. S. Huxley, A. C. Hardy, and E. B. Ford, Allen & Unwin.

METTLER, L. E. and GREGG, T. G. (1969). *Population Genetics and Evolution*, Prentice-Hall.

MEWISSEN, D. J., DAMBLON, J. and BACQ, W. M. (1960). 'Radiosensibilité des graines d'*Andropogon*, issues de terrains uranifère et non-uranifère du Katanga.' *Bull. Inst. Agron. Gembloux*, **1**, 331–8.

MOONEY, H. A. and BILLINGS, W. D. (1961). 'Comparative physiological ecology of arctic and alpine populations of *Oxyria digyna*.' *Ecol. Monogr.* **31**, 1–29.

MØRCH, E. T. (1941). *Chondrodystrophic Dwarfs in Denmark*, Ejnar Munksgaard.

PRIME, C. T. (1960). *Lords and ladies*, Collins.

SHEPPARD, P. M. (1967). *Natural Selection and Heredity*, Hutchinson, 3rd Ed.

SNAYDON, R. W. and BRADSHAW, A. D. (1961). 'Differential response to calcium within the species *Festuca ovina* L.' *New Phylotogist*, **60**, 219–34.

SNAYDON, R. W. and BRADSHAW, A. D. (1962). 'Differences between natural populations of *Trifolium repens* in response to mineral nutrients. I. Phosphates.' *J. Exp. Bot.* **13**, 422–34.

SNAYDON, R. W. and BRADSHAW, A. D. (1969). 'Differences between natural populations of *Trifolium repens* in response to mineral nutrients. II. Calcium, magnesium and potassium.' *J. Applied Ecology*, **6**, 185–202.

STEBBINS, G. L. (1957). *Variation and Evolution in Plants*, Columbia University Press. (Reprint.)

STERN, C. (1960). *Principles of Human Genetics*, Freeman (2nd Ed.), p. 462.

THODAY, J. M. (1953). In *Evolution*, Edited by R. Brown and J. F. Danielli. Cambridge University Press.

WHITEHEAD, F. W. (1962). 'Experimental studies of the effect of wind on plant growth and anatomy. II. *Helianthus annuus*.' *New Phytologist*, **61**, 59–62.

WILKINS, D. A. (1957). 'A technique for the measurement of lead tolerance in plants.' *Nature*, **180**, 37–8.

WILKINS, D. A. (1960). 'The measurement and genetical analysis of lead tolerance in *Festuca ovina*. *Scottish Plant Breeding Station Report*, pages 85–98.

WILKINS, D. A. and LEWIS, M. C. (1969). 'An application of ordination to genecology.' *New Phytologist*, **68**, 861–71.

Index

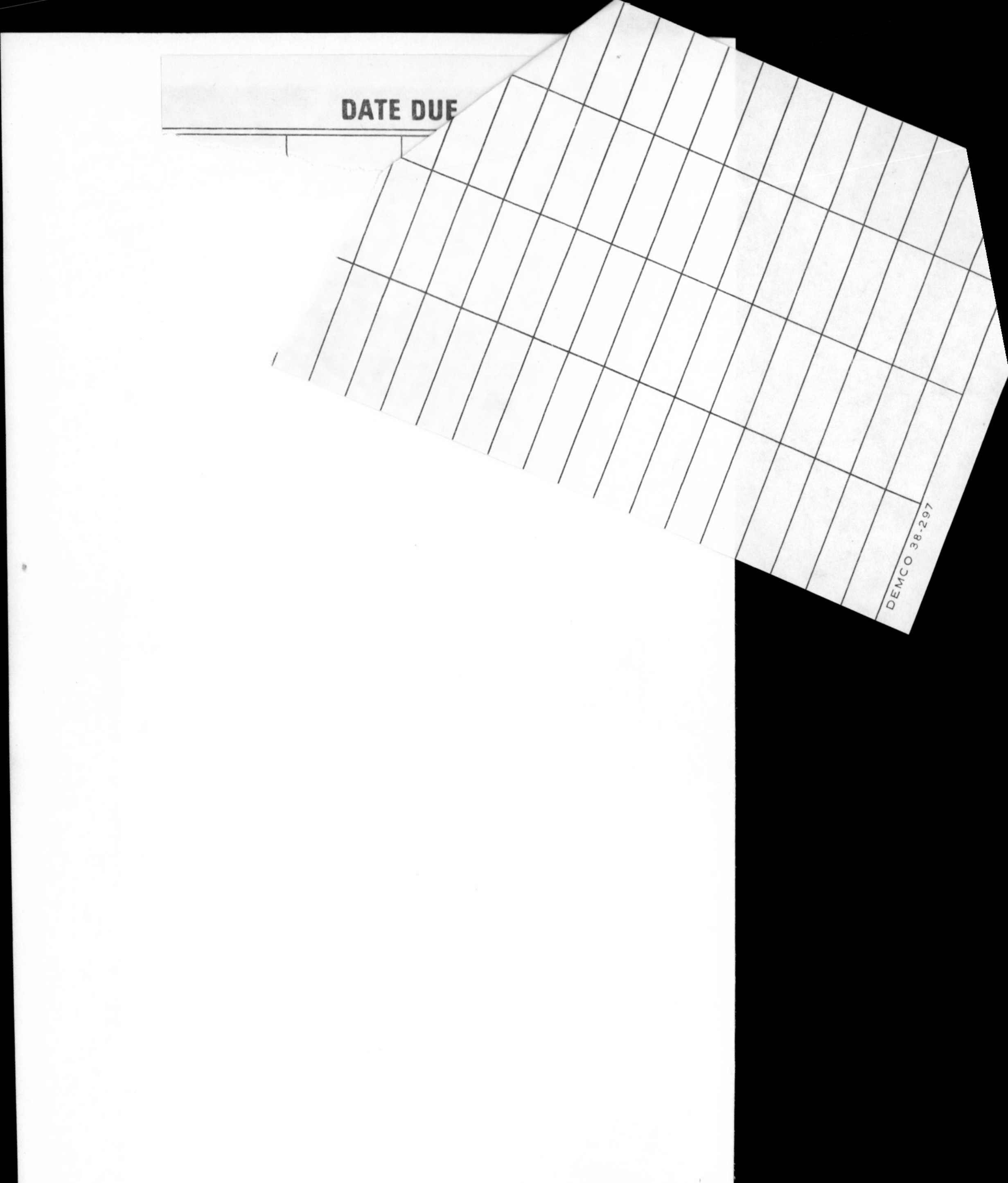